Current Topics in Environmental Health and Preventive Medicine

Series Editor

Hidekuni Inadera, Department of Public Health, University of Toyama
Toyama, Japan

Current Topics in Environmental Health and Preventive Medicine, published in partnership with the Japanese Society of Hygiene, is designed to deliver well written volumes authored by experts from around the globe, covering the prevention and environmental health related to medical, biological, molecular biological, genetic, physical, psychosocial, chemical, and other environmental factors. The series will be a valuable resource to both new and established researchers, as well as students who are seeking comprehensive information on environmental health and health promotion.

Tomoki Shiozawa • Hiromi Hirata
Takashi Inoue • Dominika Kanikowska
Hiroki Takada

Editors

Gerontology as an Interdisciplinary Science

Editors
Tomoki Shiozawa
Department of Business Administration
Aoyama Gakuin University School of
Business
Tokyo, Japan

Takashi Inoue
Department of Public and Regional
Economics
Aoyama Gakuin University, College of
Economics
Tokyo, Japan

Hiroki Takada
Department of Human and Artificial
Intelligence Systems
University of Fukui, Graduate School of
Engineering
Fukui, Fukui, Japan

Hiromi Hirata
Department of Chemistry and
Biological Science
Aoyama Gakuin University, College of
Science and Engineering
Sagamihara, Kanagawa, Japan

Dominika Kanikowska
Department of Pathophysiology
Poznan University of Medical Sciences
Poznan, Poland

ISSN 2364-8333　　　　　ISSN 2364-8341　(electronic)
Current Topics in Environmental Health and Preventive Medicine
ISBN 978-981-97-2714-8　　　ISBN 978-981-97-2712-4　(eBook)
https://doi.org/10.1007/978-981-97-2712-4

© The Editor(s) (if applicable) and The Author(s), under exclusive license to Springer Nature Singapore Pte Ltd. 2024, corrected publication 2024
This work is subject to copyright. All rights are solely and exclusively licensed by the Publisher, whether the whole or part of the material is concerned, specifically the rights of translation, reprinting, reuse of illustrations, recitation, broadcasting, reproduction on microfilms or in any other physical way, and transmission or information storage and retrieval, electronic adaptation, computer software, or by similar or dissimilar methodology now known or hereafter developed.
The use of general descriptive names, registered names, trademarks, service marks, etc. in this publication does not imply, even in the absence of a specific statement, that such names are exempt from the relevant protective laws and regulations and therefore free for general use.
The publisher, the authors and the editors are safe to assume that the advice and information in this book are believed to be true and accurate at the date of publication. Neither the publisher nor the authors or the editors give a warranty, expressed or implied, with respect to the material contained herein or for any errors or omissions that may have been made. The publisher remains neutral with regard to jurisdictional claims in published maps and institutional affiliations.

This Springer imprint is published by the registered company Springer Nature Singapore Pte Ltd.
The registered company address is: 152 Beach Road, #21-01/04 Gateway East, Singapore 189721, Singapore

If disposing of this product, please recycle the paper.

Preface

The society is experiencing significant demographic transformations characterized by phenomena including aging and increased longevity. The rising population of those reaching the age of 100 and the provision of support by society and corporations for their well-being have emerged as crucial concerns. Japan's trajectory toward a super-aging society is particularly striking. Its population has reached an aging rate of 29.1% by 2022, accompanied by an unprecedented average life expectancy of 81.47 years for males and 87.57 years for females. Gerontology has garnered considerable attention as a field endeavoring to illuminate this pressing issue and to reconceptualize the elderly as a novel sociocultural resource.

Deriving its name from the longevity research conducted by microbiologist Ilya Metchnikoff (recipient of the Nobel Prize in Physiology or Medicine) at the Pasteur Institute in France in 1903, gerontology experienced significant advancements primarily in the United States after the 1930s. Meanwhile in Japan, academic societies, such as the Japan Gerontological Society, have been actively engaged in this domain since the 1960s. In recent years, gerontology has been recognized as an interdisciplinary research field. It investigates the multifaceted aspects of aging that encompass biological, social-scientific, and psychological dimensions, thereby making valuable contributions to societies grappling with the challenges posed by super-aging.

Considering these backgrounds, we are delighted to present "Gerontology as an Interdisciplinary Science," a collaborative publication between the Japanese Society for Hygiene, Poznan Medical University, Poznan, Poland, and the Institute of Gerontology at Aoyama Gakuin University, Tokyo, Japan. This book features contributions from 18 authors, including esteemed university professors and experts in the field. It is organized into three sections, each featuring six authors, collectively encompassing a broad spectrum of gerontology-related topics. The papers of the initial section, titled "Environmental Health and Gerontology," investigate the intricate interplay between environmental factors and the aging process. The subsequent section, "Aging Societies and Gerontology," delves into the multifaceted challenges and opportunities arising from an increasingly aging population. The last section, "Preventive Medicine and Gerontology," delves into the exploration of strategies aimed at averting or postponing age-related ailments and disorders. Each of these sections offers user-friendly algorithms and key insights that facilitate comprehension. By amalgamating social science research with preventive and environmental medicine investigations, a synergistic interdisciplinary framework emerges. This engenders novel academic disciplines and propels groundbreaking innovations.

The papers featured in this book encompass a diverse range of subjects, spanning medical and physiological investigations concerning both humans and laboratory animals, as well as disciplines such as engineering, economics, and ethics. These contributions offer invaluable insights into the intricate and multifaceted dimensions of aging and its far-reaching impact on individuals, families, and societies at large. With its comprehensive coverage, this book serves as an indispensable resource for scholars specializing in environmental and occupational health worldwide, as well as for both aspiring and established researchers, students, and field practitioners; a thorough understanding of the caregiving, well-being, and health promotion aspects pertaining to older adults can be elucidated. Additionally, this book holds significance for researchers working across various fields, including robot operation interfaces and the analysis of emotional responses through biosignals among the elderly. Moreover, it serves as a valuable source of knowledge for creators and technical personnel involved in biosignal data science and database management, leading to a further step in comprehending the unique needs and characteristics of the elderly population.

Lastly, we wish to extend our heartfelt gratitude to the production editors and the editorial team at Springer Nature for their invaluable assistance in bringing this book to fruition. Their steadfast patience and unwavering commitment to professionalism have been instrumental throughout the entire editorial journey. Without their dedicated support, the publication of this book would not have been achievable.

Tokyo, Japan	Tomoki Shiozawa
Sagamihara, Kanagawa, Japan	Hiromi Hirata
Tokyo, Japan	Takashi Inoue
Poznan, Poland	Dominika Kanikowska
Fukui, Fukui, Japan	Hiroki Takada

The original version of the book has been revised. Abstracts and keywords has been updated for all chapters. The correction to this book is available at https://doi.org/10.1007/978-981-97-2712-4_19

Contents

Part I Environmental Health and Gerontology

1. **Aging and Senescence Studies in Human and Zebrafish** 3
   Hiromi Hirata, Tsuyoshi Tezuka, and Kota Ujibe

2. **Unconstrained Biosignal Measurement Technology and Its Applications in Assisting Aged Society** 23
   Yosuke Kurihara, Yuri Hamada, and Takashi Kaburagi

3. **Effects of Aging on the Vestibulo-Cardiovascular Reflex** 45
   Kunihiko Tanaka

4. **Aging and Biological Rhythms** 57
   Dominika Kanikowska

5. **Development of 3D Contents for Extracting Aging Characteristics Using Electrical Physiology** 67
   Fumiya Kinoshita, Kikuo Ito, and Hiroki Takada

6. **Comparison of Radial Motions in the Young with Those in the Elderly While Viewing 3D Video Clips Using Artificial Intelligence** 77
   Yoshiki Itatu, Hiroki Takada, and Tomoki Shiozawa

Part II Aging Society and Gerontology

7. **Elderly Migration in Tokyo Metropolitan Area, Japan** 91
   Makoto Hirai

8. **Simplified Projection of the Insurance Premiums in the Greater Tokyo Area, 2020–2060** 105
   Nozomu Inoue

| 9 | Advance Care Planning from Clinical Ethics Perspectives in Japan | 125 |

Kei Takeshita

| 10 | Social Capital Well-Being in a Super-Aging Society | 135 |

Masumi Takada and Nobuko Miyata

| 11 | Development of Medical Technology for Social Isolation in an Aging Society Through an Industry–Government–Academia Collaboration: Okumikawa Medical Valley Project | 143 |

Hidemasa Yoneda, Ryohei Hasegawa, and Hitoshi Hirata

| 12 | The Diffusion Process of Spanish Influenza Deaths in Japan: Mathematical Models Using the Gompertz and Logistic Curves | 153 |

Takashi Inoue

Part III Preventive Medicine and Gerontology

| 13 | Measuring Facial Displacements and Strains for Cosmetics Development and Beauty Care | 175 |

Satoru Yoneyama

| 14 | Educational and Collaborative Model for Early Detection and Intervention of Age-Related Hearing Loss to Enhance Health and Well-Being of the Aged | 191 |

Tomoko Sano, Noriko Katsuya, Hisao Osada, and Keiko Morita

| 15 | The Role of Aesthetics in Elderly Care | 215 |

Noriko Onishi and Toshiya Nagamatsu

| 16 | Maintenance and Improvement of Brain Health by the Brainwave-Based Brain-Training Competition, "bSports" | 225 |

Ryohei P. Hasegawa

| 17 | Wearable Systems Supporting Healthy Daily Life | 237 |

Guillaume Lopez

| 18 | Experimental Understanding of the Flow Dynamics of Exhaled Air to Prevent Infection through Aerosol | 249 |

Keiko Ishii, Yoshiko Ohno, Maiko Oikawa, and Noriko Onishi

Correction to: Gerontology as an Interdisciplinary Science.............. C1

Part I
Environmental Health and Gerontology

Chapter 1
Aging and Senescence Studies in Human and Zebrafish

Hiromi Hirata, Tsuyoshi Tezuka, and Kota Ujibe

Abstract Everyone experiences aging in his/her life. Since aging is a deteriorative process toward the end of life, most people desire longevity. However, we do not know how we can extend our life span. We do not even know what the actual cause of aging is. Senescence, which is a cellular process of aging, is a hallmark of animal aging. In this review, we introduce many papers that unveiled the process and cause of aging and senescence along with aging-related diseases in humans. We also discuss fish models to study aging and senescence.

Keywords Aging · Senescence · Cell cycle · DNA damage · Telomere · Disease · Zebrafish

Abbreviations

Aβ	Amyloid-β
CDK	Cyclin-dependent kinase
cDNA	Complementary DNA
DNA	Deoxyribonucleic acid
GFP	Green fluorescent protein
GWAS	Genome-wide associate studies
mRNA	Messenger ribonucleic acid

H. Hirata (✉) · T. Tezuka · K. Ujibe
Department of Chemistry and Biological Science, College of Science and Engineering, Aoyama Gakuin University, Sagamihara, Kanagawa, Japan
e-mail: hihirata@chem.aoyama.ac.jp

© The Author(s), under exclusive license to Springer Nature Singapore Pte Ltd. 2024, corrected publication 2024
T. Shiozawa et al. (eds.), *Gerontology as an Interdisciplinary Science*, Current Topics in Environmental Health and Preventive Medicine,
https://doi.org/10.1007/978-981-97-2712-4_1

PET	Positron emission tomography
RFP	Red fluorescent protein
RNA	Ribonucleic acid
ROS	Reactive oxygen species
RT-PCR	Reverse transcription polymerase chain reaction
SAβGal	Senescence-associated beta-galactosidase
TILLING	Targeting-induced local lesion in genome

1.1 Introduction

Many Western countries and East Asian countries undergo population aging. Indeed, the percentage of people aged 65 and over in Japan was 17% in 2000 and 28% in 2020 and will reach 35% in 2040 and 38% in 2060 (https://www8.cao.go.jp/kourei/whitepaper/index-w.html). The increase in the number of elderly people has been accelerated by reduced birth rate and extended longevity, the latter being enhanced by the recent significant advances in life science to study aging and senescence. Aging is the gradual process of becoming old that accompanies a progressive decline of physiological integrity and function. The irreversible deterioration of organs and tissues is followed by various aging-related pathology and diseases such as high blood pressure, arteriosclerosis, rheumatism, diabetes, cancer, infarction, stroke, macular degeneration, Alzheimer's disease, Parkinson's disease and many others. Senescence is another term for aging, mostly used at the cellular level. Senescent cells keep normal homeostasis and are functionally active but show cell cycle arrest and cease proliferation. By searching for scientific papers in PubMed, which is a free search engine to access publications on life science and biomedical topics, I could find several papers harboring either "aging"or "senescence"as a keyword (Fig. 1.1). The number of papers searchable by either of these keywords significantly increased in the twenty-first century, revealing that aging and senescence have been expansively explored as scientific interests.

The two major questions of aging and senescence in humans are "How do we age?" and "Why do we age?" The former is an issue of hallmarks of aging, including altered homeostasis, physiological symptoms and pathology. The latter is a mechanistic issue toward understanding the causes of aging. It is now known that senescence contributes to aging. Many genetic mutations that accelerate cellular senescence and/or animal aging have provided molecular insights into aging, leading to understanding normal homeostasis and physiological integrity that avoids senescence and aging. This review focuses on the molecular mechanisms and theories of senescence and aging in humans and mice. I also raise an emerging model animal, a zebrafish, to extensively study senescence and aging.

1 Aging and Senescence Studies in Human and Zebrafish

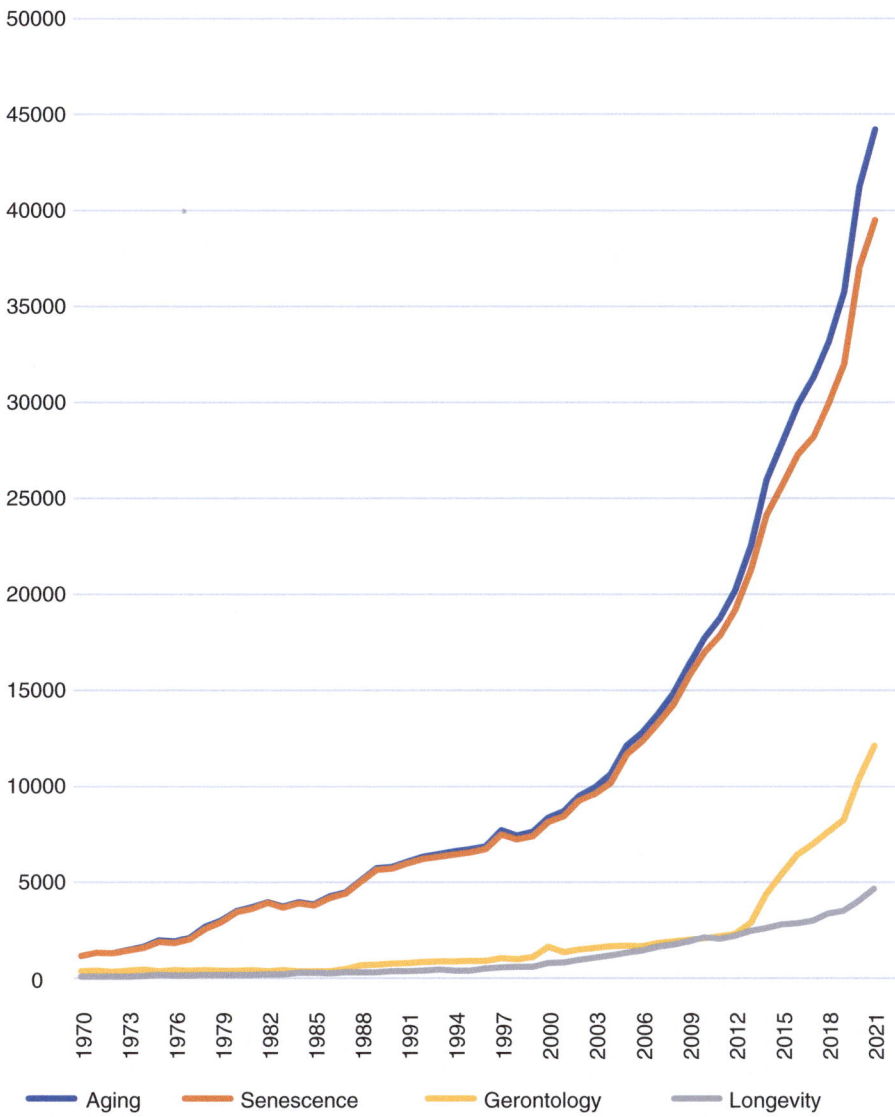

Fig. 1.1 Number of publications related to aging. A number of aging-related papers were investigated by PubMed search from 1971 to 2022

1.2 Senescence

Cellular senescence was first described in cultured human cells by Hayflick and Moorhead [1]. Normal human fetal fibroblasts cease to divide after 50 times of passage. They suggested that cells repeat cell division up to certain times, referred to as the Hayflick limit, but then enter the phase of irreversible growth arrest. The key question of this cellular senescence is how cells count the number of their cell divisions. In other words, what molecules accumulate or are depleted in the senescent cells? There are three major molecules known to regulate the cellular senescence.

1.2.1 Telomere

Telomere is a DNA-protein structure comprising repetitive hexanucleotide (TTAGGG) sequences [2, 3]. Telomere-associated shelterin proteins are located at the ends of chromosomes [4]. Although this terminal DNA region is maintained by the telomerase reverse transcriptase enzyme complex during the DNA replication process, the actual length of the telomere progressively shortens with each cell cycle division. In fact, the length of telomere shortens from 11 kb at birth to 4 kb at old [5, 6]. The shortening of telomere triggers cellular senescence and unwanted chromosome fusions [7, 8]. In agreement with these observations, cancer cells that divide continuously maintain high levels of telomerase activity [9]. Mutations in the telomerase reverse transcriptase gene cause a premature aging syndrome, dyskeratosis congenita [6]. Similarly, mutations of telomeric DNA-binding protein 1 (TRF1)-interacting nuclear factor 2 cause the same disease [10]. Mice deficient in the telomerase reverse transcriptase gene showed a short length of telomeric DNA and a short life span [11]. Conversely, mice that overexpress telomerase reverse transcriptase showed an increase in lifespan [12]. Telomere shortening is one of the major factors for cellular senescence and aging.

1.2.2 Oxidative Stress

Reactive oxygen species (ROS) is another major factor in the cellular senescence [13]. ROS is a subset of highly reactive molecules comprising superoxide anion (O_2^-), hydroxyl radical (OH^-), singlet oxygen ($^1O_2$) and hydrogen peroxide (H_2O_2). In some cases, nitric oxide (NO^-) and peroxynitrite ($ONOO^-$) are also included as ROS. Superoxide anion, which is produced by a decrease in electron transfer in mitochondrial respiration, is converted to hydrogen peroxide by superoxide dismutase (SOD). Hydrogen peroxide is normally degraded into oxygen and water by catalase. However, hydroxyl radical and singlet oxygen also arise from hydrogen peroxide at low frequencies in all cells. ROS causes unwanted oxidative reactions, including DNA base oxidation, lipid peroxidation and protein carbonylation. These damaged molecules are created continuously and repaired in normal homeostasis. In low levels of ROS, mild oxidative stress tonically activates the repair system and even contributes to lengthening the lifespans of organisms [14]. ROS is also used as a second messenger to promote cell cycle progression and cell fate decision [15, 16]. In high levels of ROS, on the other hand, severe oxidative stress causes the accumulation of DNA damage and damaged macromolecules that in turn alter gene expression and further increase ROS, especially in mitochondria, where ROS is generated. This unfortunate positive feedback changes metabolism, epigenetic regulation and homeostatic integrity at multiple steps and induces cell cycle arrest to enter the senescent state. Oxidative stress also stimulate telomere shortening [17].

1.2.3 DNA Damage

The shortening of telomere is recognized as DNA damage. In addition to the telomere shortening and ROS-mediated DNA base oxidation, spontaneous DNA mutation generated by replication errors is a major cause of DNA damage. Cells have several repair systems to cope with different types of DNA damage. The error of DNA replication and ROS-mediated base oxidation generate a mismatch of base pairs. This mismatch is repaired by either base excision repair, mismatch repair, translesion synthesis or break-induced recombination [18]. ROS also induces a single-strand break of DNA, which is repaired by a single-strand break repair. Ionizing radiation creates a double-strand break, which is the most severe damage in DNA. Double-strand break is repairable by either homologous recombination, non-homologous end joining or single-strand annealing [19]. Upon the double-strand breaks, unusual terminal ends of DNA are recognized by ataxia-telangiectasia mutated (ATM) and ataxia-telangiectasia-and-RAD3-related (ATR) proteins. ATM phosphorylates p53, which promotes stabilization and nuclear localization of p53. Activated p53 functions as a transcription factor and induces the expression of a cell cycle inhibitor gene $p21^{Cip}$. This transcriptional regulation causes cell cycle arrest to gain time to repair the damages. However, if there is too much DNA damage, p53 activation goes over the threshold, triggering the activation of the apoptosis pathway. This system eliminates DNA damage. If the cells with a number of DNA damages, including mismatch of base pairs, did not undergo apoptosis and enter the cell division, the mutation is to be fixed in DNA. Since the accumulation of DNA mutations induces cancer, DNA damage is linked to carcinogenesis. To avoid the formation of cancer cells, cells with a lot of DNA damage enter the senescent status and are sometimes removed from the tissues. Thus, the accumulation of DNA damage is a factor in cellular senescence and aging.

Phosphorylation of H2AX, a histone variant, is a marker of DNA damage. Histon octamers bind to DNA to form chromosomes and are composed of two H2A/H2B dimers, two H3 and two H4 proteins. H2AX is a minor variant of H2A [20]. In response to the double-strand breaks of DNA, the serine-139 residue at the C-terminus of H2AX is phosphorylated by ATM near the damaged sites. This phosphorylated form of H2AX is referred to as γH2AX. The γH2AX contributes to recruiting various DNA repair systems to the lesion sites. In fact, DNA damage is not repaired in H2AX-deficient cells, suggesting the essential role of γH2AX in the recognition and repair of DNA damage [21, 22]. The γH2AX protein also activates cell cycle inhibitors to induce cell cycle arrest. The amounts of γH2AX increase in senescent cells, probably because aged cells have many genomic lesions. In addition, DNA damages that remain unrepaired accumulate in senescent cells [23]. The increase of γH2AX is also seen in tissues of aged animals [24, 25], further supporting the notion that γH2AX is a marker of DNA damage as well as cellular senescence and aging.

1.2.4 Cell Cycle Inhibitors

Cell cycle inhibitors such as $p16^{INK4a}$, $p14^{ARF}$ ($p19^{ARF}$ in mice), $p21^{Cip}$ and $p15^{INK4b}$ interfere with cyclin-CDK function and cause cell cycle arrest [26]. These proteins are known as senescence markers, as they are expressed at high levels in senescent cells. The p16INK4a/p14ARF gene locus encodes p16INK4a and p14ARF proteins in a complicated manner. The transcripts of these two products are generated from different exon 1 (exon 1a and 1b) using different promoters. But they share the same exons 2 and 3. Interestingly, two transcripts use different reading frames for exons 2 and 3. Thus, $p16^{INK4a}$ and $p14^{ARF}$ do not have any amino acid homology and function in different manners. The $p16^{INK4a}$ binds to cyclin-dependent kinases CDK4 and CDK6, inhibiting CyclinD/CDK4 and Cyclin D/CDK6 function. Since these two cyclin-CDK complexes work for cell cycle progression in the G1 phase, $p16^{INK4a}$ causes cell cycle arrest at the G1 phase. The $p14^{ARF}$ binds to and inhibits MDM2, which binds to p53 and promotes ubiquitination and proteasome-dependent degradation of p53. Thus, $p14^{ARF}$ activates p53 that leads to the expression of $p21^{Cip}$, which binds to CDK4, CDK6 and CDK2 and inhibits CyclinD/CDK4, Cyclin D/CDK6 and CyclinE/CDK2 function, ceasing the cell cycle progression in G1 phase. Likewise, the $p15^{INK4b}$ protein binds to CDK4 and CDK6, contributing to the cell cycle arrest. The expression of the above cell cycle inhibitors is maintained at low levels in normal cells and tissues in young animals but becomes upregulated during continuous culture passages and animal aging [27]. Although numerous transcriptional activation, repression, and epigenetic regulations of these genes have been reported in the context of cancer development, transcriptional regulation of these genes in senescence remains to be solved [28]. Among these cell cycle inhibitors, $p16^{INK4a}$ has been suggested to be most relevant to senescence. The expression of $p16^{INK4a}$ increases gradually during development, keeping the low level in adulthood, but becomes dramatically upregulated in the old animals [29]. This significant increase in $p16^{INK4a}$ expression in aged animals suggests that $p16^{INK4a}$ is the most prominent marker of aging and senescence [30].

1.2.5 Senescence-Associated β-Galactosidase

Senescence-associated beta-galactosidase (SAβGal) is a classic marker of cellular senescence originally reported in 1995 [31]. Generally, β-galactosidase activity is assayed by the hydrolysis of 5-bromo-4-chloro-3-indolyl-β-D-galactoside (X-gal) into monosaccharides. This product forms dark blue precipitates after oxidation, enabling the visualization of β-galactosidase activity as the blue color. The SAβGal, which was detectable specifically at pH 6 as blue labeling, was upregulated in senescent cells and aged tissues but not in young cells, terminally differentiated cells and immortal cells as well as in young tissues. SAβGal is a lysosomal β-galactosidase encoded by *galactosidase beta 1 (GLB1)* gene [32]. Since SAβGal is an easy assay for detecting the distribution of aged cells, this evaluation method has been widely used for cellular senescence and animal aging research.

1.3 Aging

Aging is an intrinsic functional decline of homeostasis and physiological integrity [33]. What is the direct cause of aging remains a challenging question. Undoubtedly, cellular senescence is one of the major direct causes of aging. However, several genes induce senescence and harbor anti-senescent functions such as apoptosis and quiescence. This paradoxical function makes it difficult to clarify the relationship between senescence and aging. There are a number of aging-related diseases in humans, as follows.

1.3.1 Cancer

Cancer is characterized by abnormal cell proliferation, invasion from the tissue and metastasis to the other tissues. Cancer is the second leading cause of death globally. Indeed, 27% of the cause of death in Japan was cancer in 2021. Since cancer is induced by DNA mutations and since tumor suppressor genes such as p53 and $p16^{INK4a}/p14^{ARF}$ are involved in the induction of cellular senescence, senescence is likely a tumor suppressor mechanism and thus cancer is a failure of the cellular senescence [34, 35]. Similarly, the increase of senescent cells with age increases cancer incidence. In agreement with this notion, if senescent cells are removed from the aged tissues, the onset of tumor formation is delayed [36]. This strategy is potentially applicable to cancer elimination, which is referred to as senolytic therapy [37].

1.3.2 Cardiovascular Disease

Cardiovascular diseases are disorders of the heart and blood vessels and are the leading cause of death globally. The blood vessels of the cardiac muscles, brain and arms/legs are affected by coronary heart disease, cerebrovascular disease and peripheral arterial disease, respectively. The risk of arteriosclerosis of any artery blood vessels increases with age, which in turn causes rupture and infarction of blood vessels, leading to cardiovascular disease. Macrophage cells are the primary senescent cells that show high SAβGal activities in the blood vessels [38]. Cardiomyocyte atrophy is another infarction disease that frequently occurs in old people [39].

1.3.3 Kidney Dysfunction

The glomerular filtration rate decreases and urine albumin increases with age through the development of nephrosclerosis and nephron atrophy. Cellular senescence marked by increased p53 and p16^{INK4a} expression is linked with these kidney aging [40–42].

1.3.4 Diabetes

Diabetes is a hyperglycemia disease, which is also evident in the high glucose in the urine [43]. Type 1 diabetes occurs by the elimination of the insulin-producing Langerhans β cells in the pancreas. Insulin is a hormone that reduces blood sugar levels. In many cases of type 1 diabetes, the immune system attacks and eliminates the β cells. Type 2 diabetes occurs due to the insufficient use of insulin. More than 90% of the patients who have type 2 diabetes are deteriorated by excess body weight and physical inactivity along with age. Genome-wide associate studies (GWAS) have revealed that SNPs in the p16^{INK4a}/p14^{ARF} locus are linked to the risk of type 2 diabetes [44, 45]. This evidence suggests that cellular senescence causes diabetes as an age-associated disease.

1.3.5 Cirrhosis

Cirrhosis is a fibrosis of the liver caused by long-term liver damage, including nonalcoholic fatty liver disease, hepatitis and excess alcohol consumption [46]. The senescence of hepatocytes is also linked to nonalcoholic liver disorders [47, 48]. Senolytic elimination of senescent hepatocytes improved liver condition and suppressed fat accumulation [49].

1.3.6 Sarcopenia

Sarcopenia is a muscle atrophy that deteriorates with age and physical inactivity [50]. Skeletal muscle cells can be classified into fast-twitch and slow-twitch muscle fibers. While fast-twitch muscles make sudden and powerful movements in sprints, slow-twitch muscles enable sustained movements in the marathon. In young individuals, the decrease of muscle is compensated by myogenesis, in which muscle stem cells, which are often referred to as muscle satellite cells, differentiate into muscle fibers [51]. This regeneration potential for both fast-twitch fibers and slow-twitch fibers declines with age, along with the accumulation of p16^{INK4a} in muscle satellite cells [52]. In sarcopenia, the reduction of muscle fibers occurs predominantly in fast-twitch muscles. Interestingly, senolytic elimination of senescent muscle satellite cells enhanced myogenesis [53].

1.3.7 Osteoporosis

Bones are continuously destructed by osteoclasts and built by osteoblasts, and thus, they are kept remodeling for a lifetime [54]. But as we get older, we lose more bone than we build due to the dysregulation of bone homeostasis. Osteoporosis is the reduction of bone density with aging that causes bone fractures. These osteoporotic fractures restrict movements and then promote sarcopenia, thus triggering positive feedback toward bedridden or immotile life. In aged bones, senescent osteoblasts that express high levels of p16^{INK4a}, p21^{Cip} and p53 and show cell cycle arrest were found [55]. Targeted removal of senescent osteoblasts mitigated osteoporosis symptoms in aged mice [56].

1.3.8 Neurodegenerative Diseases

Neurodegenerative diseases are disorders caused by progressive loss of neuronal cells [57]. The major causes of neuronal cell death are oxidative stress, inflammation, neurotoxins, insufficient neuronal excitability, poor nutrition, mitochondrial dysfunction and accumulation of abnormal proteins, triggering the apoptosis pathways. In the world. About 50 million people currently suffer from neurodegenerative diseases, and the symptoms deteriorate with age. It is suggested that the number of neurodegenerative disease patients will increase to 115 million by 2050, along with the increase in our lifespan [58].

Alzheimer's disease is the most common dementia characterized by progressive loss of brain functions such as memory, language, thinking, mental control and consciousness [59]. In patients, the shrinkage of the brain can be observed by positron emission tomography (PET), especially in the hippocampal area, which is a brain region for memory storage and retrieval. Abnormal aggregation of the amyloid-β (Aβ) peptides and tau proteins in the brain is the major pathological condition and one of the direct causes of Alzheimer's disease. However, the actual cause for most patients is unclear. About 1–2% of Alzheimer's disease are inherited in an autosomal dominant manner [60]. Mutations in either Aβ precursor protein, presenilin 1 or presenilin 2, the latter two promoting the aggregation of Aβ, are responsible for the genetic cases of Alzheimer's disease [61]. Although our brain can remove these protein aggregations, the capability of the clearance decreases with age. Thus, age-dependent accumulation of harmful proteins causes Alzheimer's disease in old people.

Parkinson's disease is a neurodegenerative disorder affecting the motor system and mental control [62]. The typical symptom of this disease is motor disability, including tremors, stiffness and slow movement. In patients, dopaminergic neurons in the substantia nigra of the midbrain are eliminated along with age, triggering the motor and mood deficits. Cytoplasmic and intranuclear accumulation of a-synuclein protein are typical features and one of the direct causes of Parkinson's disease [63]. The clearance capability of these abnormal protein aggregates decreases with age. About 5–10% of Parkinson's disease are inherited in an autosomal dominant or recessive manner. One of the causative genes is *SNCA*, encoding the α-synuclein, which forms cytotoxic aggregates that kill dopaminergic neurons [64].

Macular degeneration is another neurodegenerative disorder characterized by the loss of the macula of the retina in the eye [65]. Macula is located at the center of the retina and is necessary to collect the central vision. Macular degeneration patients gradually lose the middle part rather than the peripheral part of the vision. About 50% of macular degeneration is linked to several chromosome loci [66]. Indeed, some genes encoding for complement factor proteins are associated with the risk, revealing that inflammation is pathologically relevant for macular degeneration [67]. The accumulation of intracellular and extracellular debris such as lysosomal lipofuscin, retinal drusen and metabolic wastes generated by the inflammation deteriorates age-related macular degeneration.

Taken together, the above neurodegenerative diseases have similar pathological bases related to the toxic aggregates. First, abnormal protein aggregates inside and outside of neurons are formed. Second, the protein aggregates accumulate with age. Finally, the aggregates promote cell death in the specific neurons/neuronal tissues and develop age-related neuronal disorders.

1.4 Fish Models

Model animals have been applied for biomedical studies including aging and senescence. Among a variety of model animals, freshwater fishes such as zebrafish and medaka have been extensively used in the twenty-first century. Zebrafish became the second major model vertebrate animal next to mice. Indeed, the number of hits by PubMed search using "zebrafish" as a keyword is now larger than those using "Drosophila," "*C. elegans*" or "*Saccharomyces cerevisiae*" (Fig. 1.2).

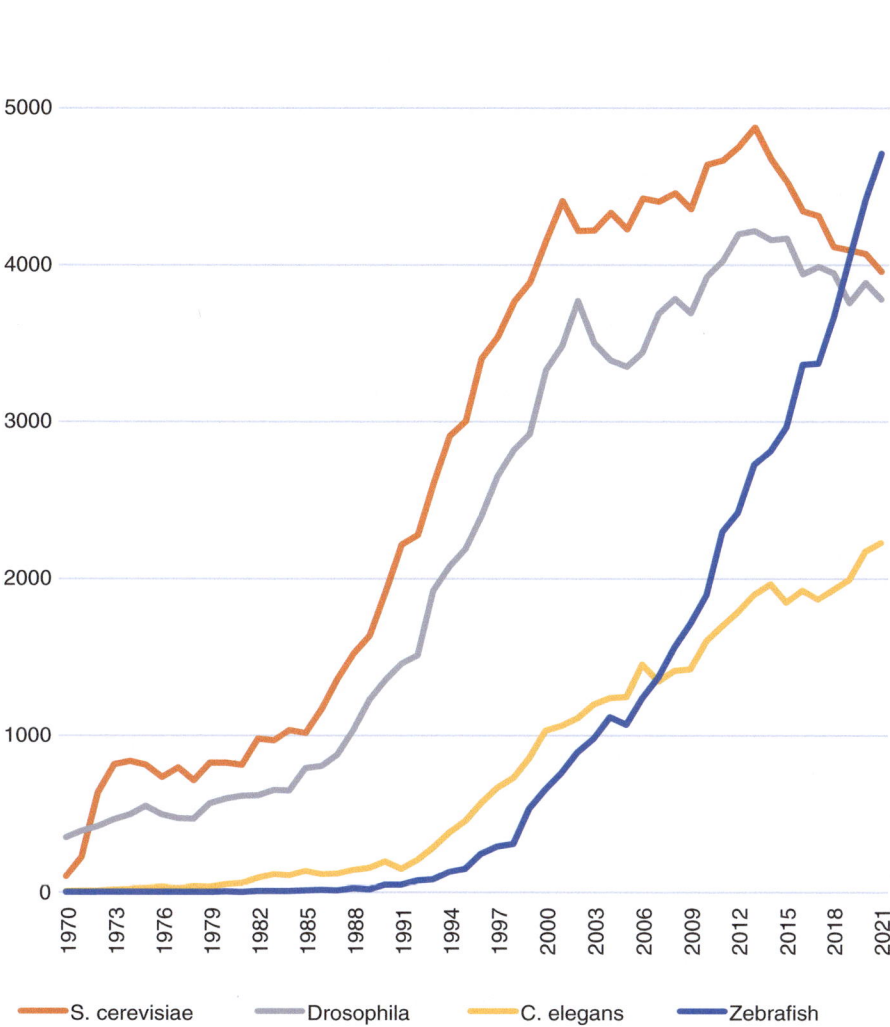

Fig. 1.2 Number of publications related to model organisms. A number of model organism-related papers were investigated by PubMed search from 1971 to 2022

1.4.1 Advantages of Zebrafish

Zebrafish (*Danio rerio*) have many advantages in biomedical research [68]. First, the husbandry of zebrafish is space- and cost-effective than that of mice. A pair of adult zebrafish can produce 200 fertilized eggs every week. They become sexually mature 3 months after fertilization. Second, the development of zebrafish takes place externally and rapidly, with the speed of embryogenesis 80-fold faster than humans [69]. Indeed, zebrafish hatches from chorion within 3 days of development, while the human gestation period is 266 days. Third, methods of genetic manipulation have been established in zebrafish to generate mutants by forward and reverse genetics as well as transgenics by transposon system [70–82]. Several mutagenesis projects have created a number of zebrafish mutants since the 1990s [83–85]. In conjunction with the transparency during embryogenesis, tissue morphogenesis can be visualized in live zebrafish. These advantages make the zebrafish a powerful vertebrate model for studying development, disease and aging [86].

1.4.2 Senescence and Aging Studies in Zebrafish

In zebrafish, cellular senescence occurs, just as in mammals [87–89]. The increase of SAβGal activities in aged fish's skin, muscle and retinal cells has been reported [90]. Expression of cell cycle inhibitors such as $p16^{INK4a}$, $p15^{INK4b}$ and $p21^{CIP}$ accompanies the nuclear accumulation of γH2AX [91].

The lifespan of zebrafish is 3–4 years, while some fish survive over 5 years under a good husbandry condition [90, 92]. The curvature of the spine and thus hunchbacked appearance is often seen in aged zebrafish. Along with age, they also show an increase in SAβGal activities in the skin and lipofuscin, which are pigment granules in the liver. Zebrafish also show a decline in regeneration capacity and motor ability [93, 94]. Telomere-deficient zebrafish showed increased DNA damage markers such as p53 binding protein and γH2AX as well as apoptosis [95, 96]. They also exhibited accelerated degenerative phenotypes, including emaciation, infertility, inflammation and shortened half-life (0.8–1.3 years). Interestingly, most offspring obtained from a pair of telomere-deficient fish and wild-type fish showed developmental malformations. Taken together, aging in zebrafish seems to occur as in mammals.

1.4.3 Diabetes Models in Zebrafish

Excess feeding or high-fat diet can induce obesity and diabetes symptoms in zebrafish [97, 98]. Expression of a constitutive-active variant of human AKT1 with a cell membrane-anchoring myristoylation signal in skin epidermal cells caused hypertrophic skin growth and obese phenotypes at the adult stage, the latter including the increase of body weight, hyperplastic growth of adipocytes, excess accumulation of fat tissues and impaired glucose intolerance [99]. Since these phenotypes are partly ameliorated by antisense morpholino-mediated knockdown of mTOR homolog in the transgenic zebrafish, the Akt-mTOR axis plays a conserved role in the diabetes of vertebrates [100].

1.4.4 Sarcopenia Models in Zebrafish

Zebrafish have fast-twitch and slow-twitch muscle fibers. While these two types of muscle cells are mosaically distributed in human musculature, fast-twitch and slow-twitch muscle fibers are exclusively located at deep layers and superficial areas underneath the skin, respectively, in zebrafish [101]. This segregation enables easy identification of muscle cell types in zebrafish. Aged zebrafish show a reduction of muscle mass in both fast-twitch and slow-twitch muscle fibers that correlate with the deterioration of swimming performance [94]. So far, many genetic mutants that have defects in fast- and/or slow-twitch muscle development have been generated [102]. Some were established as muscular dystrophy models [103–110] and myopathy models [111–123]. However, these mutants died during embryogenesis and were not available as a sarcopenia model. Zebrafish that display adult-onset muscle atrophy are necessary to assess sarcopenia in zebrafish.

1.4.5 Neurodegenerative Diseases Models in Zebrafish

Progressive degeneration of neuronal cells has also been studied in zebrafish. Zebrafish treated with scopolamine or okadaic acids, which are antagonists of acetylcholine receptors or protein phosphatase 2A, respectively, showed behavioral and histological defects associated with Alzheimer's disease [124, 125]. Zebrafish have two amyloid peptide precursor homologs (*appa* and *appb*) and *appb*-deficient zebrafish showed impairments in locomotion and behavior [126, 127]. Expression of a pathogenic variant of amyloid peptide precursor under the *appb* promoter in zebrafish generated an Alzheimer's disease, in which behavioral deficits, beta amyloidosis and neuronal cell death are observed [128].

In zebrafish, administration of neurotoxins such as 1-Methyl-4-phenyl-1,2,3,6-tetrahydropyridine (MPTP) and rotenone-induced cell death of dopaminergic neurons in the midbrain. These treatments generate models of Parkinson's disease [129–131]. These disease models show both motor deficits and non-motor behavioral impairments, which are characteristic of Parkinson's disease in humans. Genome-wide association studies in humans have identified many genes linked to Parkinson's disease. Many zebrafish models of Parkinson's disease have been generated by knockdown, knockout and transgenic expression [132–140]. For example, *park2* knockdown zebrafish showed reduced dopaminergic neurons and impaired mitochondrial respiratory functions [141, 142]. Similarly, *pink1*-deficient zebrafish displayed the loss of dopaminergic neurons and dysfunction of mitochondria [143, 144]. Other than neurodegenerative ones, there are many neurologically defective mutants in zebrafish [145–149].

1.5 Concluding Remarks

Aging and senescence are undoubtedly some of the biggest issues in life science, but they are still thoroughly unexplored. Recent advances in genome sequencing, genome editing and in vivo analyses have expanded our knowledge on the process of aging and senescence. Furthermore, animal models, including emerging zebrafish models, became available for manipulations and detailed analyses of the basis of aging and senescence as well as therapy experiments and drug screening. Future studies of these non-human studies will enable practical approaches for anti-aging.

Acknowledgments We apologize to investigators whose work could not be cited in this manuscript owing to space limitations. This work was supported by the Grant-in-Aid for Scientific Research B (19H03329) from the Ministry of Education, Culture, Sports, Science and Technology (MEXT), Japan, the Takeda Science Foundation, the Naito Foundation, and the Long-Range Research Initiative Grant from the Japan Chemical Industry Association.

Conflict of Interest The authors declare that the research was conducted in the absence of any commercial or financial relationships that could be construed as a potential conflict of interest.

Author Contributions H. H., T. T. and U. K. performed search and analyzed data; H. H. wrote the manuscript.

References

1. Hayflick L, Moorhead PS. The serial cultivation of human diploid cell strains. Exp Cell Res. 1961;25:585–621.
2. Creighton HB, McClintock B. A correlation of cytological and genetical crossing-over in Zea Mays. Proc Natl Acad Sci U S A. 1931;17(8):492–7.
3. Revy P, Kannengiesser C, Bertuch AA. Genetics of human telomere biology disorders. Nat Rev Genet. 2023;24(2):86–108.
4. de Lange T. Shelterin-mediated telomere protection. Annu Rev Genet. 2018;52:223–47.
5. Okuda K, et al. Telomere length in the newborn. Pediatr Res. 2002;52(3):377–81.
6. Bischoff C, et al. Telomere length among the elderly and oldest-old. Twin Res Hum Genet. 2005;8(5):425–32.
7. Palm W, de Lange T. How shelterin protects mammalian telomeres. Annu Rev Genet. 2008;42:301–34.
8. Barnes PJ. Mechanisms of development of multimorbidity in the elderly. Eur Respir J. 2015;45(3):790–806.
9. Stewart SA, Weinberg RA. Telomeres: cancer to human aging. Annu Rev Cell Dev Biol. 2006;22:531–57.
10. Walne AJ, et al. TINF2 mutations result in very short telomeres: analysis of a large cohort of patients with dyskeratosis congenita and related bone marrow failure syndromes. Blood. 2008;112(9):3594–600.
11. Blasco MA, et al. Telomere shortening and tumor formation by mouse cells lacking telomerase RNA. Cell. 1997;91(1):25–34.

12. Gonzalez-Suarez E, et al. Antagonistic effects of telomerase on cancer and aging in K5-mTert transgenic mice. Oncogene. 2005;24(13):2256–70.
13. Finkel T, Holbrook NJ. Oxidants, oxidative stress and the biology of ageing. Nature. 2000;408(6809):239–47.
14. Ristow M, Schmeisser S. Extending life span by increasing oxidative stress. Free Radic Biol Med. 2011;51(2):327–36.
15. Verbon EH, Post JA, Boonstra J. The influence of reactive oxygen species on cell cycle progression in mammalian cells. Gene. 2012;511(1):1–6.
16. Kaminskyy VO, Zhivotovsky B. Free radicals in cross talk between autophagy and apoptosis. Antioxid Redox Signal. 2014;21(1):86–102.
17. Passos JF, Saretzki G, von Zglinicki T. DNA damage in telomeres and mitochondria during cellular senescence: is there a connection? Nucleic Acids Res. 2007;35(22):7505–13.
18. Clarke TL, Mostoslavsky R. DNA repair as a shared hallmark in cancer and ageing. Mol Oncol. 2022;16(18):3352–79.
19. Waterman DP, Haber JE, Smolka MB. Checkpoint responses to DNA double-strand breaks. Annu Rev Biochem. 2020;89:103–33.
20. Rogakou EP, et al. DNA double-stranded breaks induce histone H2AX phosphorylation on serine 139. J Biol Chem. 1998;273(10):5858–68.
21. Celeste A, et al. Histone H2AX phosphorylation is dispensable for the initial recognition of DNA breaks. Nat Cell Biol. 2003;5(7):675–9.
22. Celeste A, et al. H2AX haploinsufficiency modifies genomic stability and tumor susceptibility. Cell. 2003;114(3):371–83.
23. Sedelnikova OA, et al. Senescing human cells and ageing mice accumulate DNA lesions with unrepairable double-strand breaks. Nat Cell Biol. 2004;6(2):168–70.
24. Bakkenist CJ, et al. Disappearance of the telomere dysfunction-induced stress response in fully senescent cells. Cancer Res. 2004;64(11):3748–52.
25. d'Adda di Fagagna F, et al. A DNA damage checkpoint response in telomere-initiated senescence. Nature. 2003;426(6963):194–8.
26. Malumbres M, Barbacid M. Cell cycle, CDKs and cancer: a changing paradigm. Nat Rev Cancer. 2009;9(3):153–66.
27. Krishnamurthy J, et al. Ink4a/Arf expression is a biomarker of aging. J Clin Invest. 2004;114(9):1299–307.
28. Sharpless NE, Sherr CJ. Forging a signature of in vivo senescence. Nat Rev Cancer. 2015;15(7):397–408.
29. Wagner KD, Wagner N. The senescence markers p16INK4A, p14ARF/p19ARF, and p21 in organ development and homeostasis. Cells. 2022;11(12):1966.
30. Gonzalez-Gualda E, et al. A guide to assessing cellular senescence in vitro and in vivo. FEBS J. 2021;288(1):56–80.
31. Dimri GP, et al. A biomarker that identifies senescent human cells in culture and in aging skin in vivo. Proc Natl Acad Sci U S A. 1995;92(20):9363–7.
32. Lee BY, et al. Senescence-associated beta-galactosidase is lysosomal beta-galactosidase. Aging Cell. 2006;5(2):187–95.
33. Lopez-Otin C, et al. Hallmarks of aging: an expanding universe. Cell. 2023;186(2):243–78.
34. Krtolica A, et al. Senescent fibroblasts promote epithelial cell growth and tumorigenesis: a link between cancer and aging. Proc Natl Acad Sci U S A. 2001;98(21):12072–7.
35. Coppe JP, et al. Senescence-associated secretory phenotypes reveal cell-nonautonomous functions of oncogenic RAS and the p53 tumor suppressor. PLoS Biol. 2008;6(12):2853–68.
36. Baker DJ, et al. Naturally occurring p16(Ink4a)-positive cells shorten healthy lifespan. Nature. 2016;530(7589):184–9.
37. Chang J, et al. Clearance of senescent cells by ABT263 rejuvenates aged hematopoietic stem cells in mice. Nat Med. 2016;22(1):78–83.
38. Childs BG, et al. Senescent intimal foam cells are deleterious at all stages of atherosclerosis. Science. 2016;354(6311):472–7.

39. Niccoli T, Partridge L. Ageing as a risk factor for disease. Curr Biol. 2012;22(17):R741–52.
40. Sturmlechner I, et al. Cellular senescence in renal ageing and disease. Nat Rev Nephrol. 2017;13(2):77–89.
41. Melk A, et al. Expression of p16INK4a and other cell cycle regulator and senescence associated genes in aging human kidney. Kidney Int. 2004;65(2):510–20.
42. Melk A, et al. Cell senescence in rat kidneys in vivo increases with growth and age despite lack of telomere shortening. Kidney Int. 2003;63(6):2134–43.
43. Christ-Crain M, et al. Diabetes insipidus. Nat Rev Dis Primers. 2019;5(1):54.
44. Zeggini E, et al. Replication of genome-wide association signals in UK samples reveals risk loci for type 2 diabetes. Science. 2007;316(5829):1336–41.
45. Jeck WR, Siebold AP, Sharpless NE. Review: a meta-analysis of GWAS and age-associated diseases. Aging Cell. 2012;11(5):727–31.
46. Gines P, et al. Liver cirrhosis. Lancet. 2021;398(10308):1359–76.
47. Hardy T, et al. Nonalcoholic fatty liver disease: pathogenesis and disease Spectrum. Annu Rev Pathol. 2016;11:451–96.
48. Pellicoro A, et al. Liver fibrosis and repair: immune regulation of wound healing in a solid organ. Nat Rev Immunol. 2014;14(3):181–94.
49. Ogrodnik M, Jurk D. Senescence explains age- and obesity-related liver steatosis. Cell Stress. 2017;1(1):70–2.
50. Cruz-Jentoft AJ, Sayer AA. Sarcopenia. Lancet. 2019;393(10191):2636–46.
51. Gopinath SD, Rando TA. Stem cell review series: aging of the skeletal muscle stem cell niche. Aging Cell. 2008;7(4):590–8.
52. Sousa-Victor P, et al. Geriatric muscle stem cells switch reversible quiescence into senescence. Nature. 2014;506(7488):316–21.
53. Baker DJ, et al. Clearance of p16Ink4a-positive senescent cells delays ageing-associated disorders. Nature. 2011;479(7372):232–6.
54. Black DM, Rosen CJ. Clinical practice. Postmenopausal osteoporosis. N Engl J Med. 2016;374(3):254–62.
55. Farr JN, et al. Identification of senescent cells in the bone microenvironment. J Bone Miner Res. 2016;31(11):1920–9.
56. Farr JN, et al. Targeting cellular senescence prevents age-related bone loss in mice. Nat Med. 2017;23(9):1072–9.
57. Hou Y, et al. Ageing as a risk factor for neurodegenerative disease. Nat Rev Neurol. 2019;15(10):565–81.
58. Livingston G, et al. Dementia prevention, intervention, and care: 2020 report of the lancet commission. Lancet. 2020;396(10248):413–46.
59. Scheltens P, et al. Alzheimer's disease. Lancet. 2021;397(10284):1577–90.
60. Long JM, Holtzman DM. Alzheimer disease: an update on pathobiology and treatment strategies. Cell. 2019;179(2):312–39.
61. Atri A. Current and future treatments in Alzheimer's disease. Semin Neurol. 2019;39(2):227–40.
62. Tolosa E, et al. Challenges in the diagnosis of Parkinson's disease. Lancet Neurol. 2021;20(5):385–97.
63. Lashuel HA, et al. The many faces of alpha-synuclein: from structure and toxicity to therapeutic target. Nat Rev Neurosci. 2013;14(1):38–48.
64. Polymeropoulos MH, et al. Mutation in the alpha-synuclein gene identified in families with Parkinson's disease. Science. 1997;276(5321):2045–7.
65. Mitchell P, et al. Age-related macular degeneration. Lancet. 2018;392(10153):1147–59.
66. Schmitz-Valckenberg S, et al. Conversion from intermediate age-related macular degeneration to geographic atrophy in a Proxima B subcohort using a multimodal approach. Ophthalmologica. 2021;244(6):523–34.
67. Edwards AO, et al. Complement factor H polymorphism and age-related macular degeneration. Science. 2005;308(5720):421–4.

68. Grunwald DJ, Eisen JS. Headwaters of the zebrafish—emergence of a new model vertebrate. Nat Rev Genet. 2002;3(9):717–24.
69. Kimmel CB, et al. Stages of embryonic development of the zebrafish. Dev Dyn. 1995;203(3):253–310.
70. Amsterdam A, et al. A large-scale insertional mutagenesis screen in zebrafish. Genes Dev. 1999;13(20):2713–24.
71. Amsterdam A, et al. Identification of 315 genes essential for early zebrafish development. Proc Natl Acad Sci U S A. 2004;101(35):12792–7.
72. Doyon Y, et al. Heritable targeted gene disruption in zebrafish using designed zinc-finger nucleases. Nat Biotechnol. 2008;26(6):702–8.
73. Gaiano N, et al. Insertional mutagenesis and rapid cloning of essential genes in zebrafish. Nature. 1996;383(6603):829–32.
74. Golling G, et al. Insertional mutagenesis in zebrafish rapidly identifies genes essential for early vertebrate development. Nat Genet. 2002;31(2):135–40.
75. Hwang WY, et al. Efficient genome editing in zebrafish using a CRISPR-Cas system. Nat Biotechnol. 2013;31(3):227–9.
76. Kawakami K, Shima A, Kawakami N. Identification of a functional transposase of the Tol2 element, an ac-like element from the Japanese medaka fish, and its transposition in the zebrafish germ lineage. Proc Natl Acad Sci U S A. 2000;97(21):11403–8.
77. Kawakami K, et al. A transposon-mediated gene trap approach identifies developmentally regulated genes in zebrafish. Dev Cell. 2004;7(1):133–44.
78. Kimura Y, et al. Efficient generation of knock-in transgenic zebrafish carrying reporter/driver genes by CRISPR/Cas9-mediated genome engineering. Sci Rep. 2014;4:6545.
79. Lin S, et al. Integration and germ-line transmission of a pseudotyped retroviral vector in zebrafish. Science. 1994;265(5172):666–9.
80. Meng X, et al. Targeted gene inactivation in zebrafish using engineered zinc-finger nucleases. Nat Biotechnol. 2008;26(6):695–701.
81. Nasevicius A, Ekker SC. Effective targeted gene 'knockdown' in zebrafish. Nat Genet. 2000;26(2):216–20.
82. Wienholds E, et al. Target-selected inactivation of the zebrafish rag1 gene. Science. 2002;297(5578):99–102.
83. Driever W, et al. A genetic screen for mutations affecting embryogenesis in zebrafish. Development. 1996;123:37–46.
84. Haffter P, et al. The identification of genes with unique and essential functions in the development of the zebrafish, *Danio rerio*. Development. 1996;123:1–36.
85. Mullins MC, et al. Large-scale mutagenesis in the zebrafish: in search of genes controlling development in a vertebrate. Curr Biol. 1994;4(3):189–202.
86. MacRae CA, Peterson RT. Zebrafish as tools for drug discovery. Nat Rev Drug Discov. 2015;14(10):721–31.
87. Kishi S. The search for evolutionary developmental origins of aging in zebrafish: a novel intersection of developmental and senescence biology in the zebrafish model system. Birth Defects Res C Embryo Today. 2011;93(3):229–48.
88. Kishi S. Using zebrafish models to explore genetic and epigenetic impacts on evolutionary developmental origins of aging. Transl Res. 2014;163(2):123–35.
89. Kishi S, et al. Zebrafish as a genetic model in biological and behavioral gerontology: where development meets aging in vertebrates—a mini-review. Gerontology. 2009;55(4):430–41.
90. Kishi S, et al. The identification of zebrafish mutants showing alterations in senescence-associated biomarkers. PLoS Genet. 2008;4(8):e1000152.
91. Haraoka Y, et al. Zebrafish imaging reveals TP53 mutation switching oncogene-induced senescence from suppressor to driver in primary tumorigenesis. Nat Commun. 2022;13(1):1417.
92. Gerhard GS, et al. Life spans and senescent phenotypes in two strains of zebrafish (*Danio rerio*). Exp Gerontol. 2002;37(8–9):1055–68.

93. Tsai SB, et al. Differential effects of genotoxic stress on both concurrent body growth and gradual senescence in the adult zebrafish. Aging Cell. 2007;6(2):209–24.
94. Gilbert MJ, Zerulla TC, Tierney KB. Zebrafish (*Danio rerio*) as a model for the study of aging and exercise: physical ability and trainability decrease with age. Exp Gerontol. 2014;50:106–13.
95. Henriques CM, et al. Telomerase is required for zebrafish lifespan. PLoS Genet. 2013;9(1):e1003214.
96. Anchelin M, et al. Premature aging in telomerase-deficient zebrafish. Dis Model Mech. 2013;6(5):1101–12.
97. Oka T, et al. Diet-induced obesity in zebrafish shares common pathophysiological pathways with mammalian obesity. BMC Physiol. 2010;10:21.
98. Meguro S, Hasumura T, Hase T. Body fat accumulation in zebrafish is induced by a diet rich in fat and reduced by supplementation with green tea extract. PLoS One. 2015;10(3):e0120142.
99. Chu CY, et al. Overexpression of Akt1 enhances adipogenesis and leads to lipoma formation in zebrafish. PLoS One. 2012;7(5):e36474.
100. Saxton RA, Sabatini DM. mTOR signaling in growth, metabolism, and disease. Cell. 2017;169(2):361–71.
101. Currie PD, Ingham PW. Induction of a specific muscle cell type by a hedgehog-like protein in zebrafish. Nature. 1996;382(6590):452–5.
102. Granato M, et al. Genes controlling and mediating locomotion behavior of the zebrafish embryo and larva. Development. 1996;123:399–413.
103. Bassett DI, et al. Dystrophin is required for the formation of stable muscle attachments in the zebrafish embryo. Development. 2003;130(23):5851–60.
104. Hall TE, et al. The zebrafish candyfloss mutant implicates extracellular matrix adhesion failure in laminin alpha2-deficient congenital muscular dystrophy. Proc Natl Acad Sci U S A. 2007;104(17):7092–7.
105. Liang WC, et al. Congenital muscular dystrophy with fatty liver and infantile-onset cataract caused by TRAPPC11 mutations: broadening of the phenotype. Skelet Muscle. 2015;5:29.
106. Schindler RF, et al. POPDC1(S201F) causes muscular dystrophy and arrhythmia by affecting protein trafficking. J Clin Invest. 2016;126(1):239–53.
107. Steffen LS, et al. The zebrafish runzel muscular dystrophy is linked to the titin gene. Dev Biol. 2007;309(2):180–92.
108. Ono F, et al. Paralytic zebrafish lacking acetylcholine receptors fail to localize rapsyn clusters to the synapse. J Neurosci. 2001;21(15):5439–48.
109. Ono F, Mandel G, Brehm P. Acetylcholine receptors direct rapsyn clusters to the neuromuscular synapse in zebrafish. J Neurosci. 2004;24(24):5475–81.
110. Ono F, et al. The zebrafish motility mutant twitch once reveals new roles for rapsyn in synaptic function. J Neurosci. 2002;22(15):6491–8.
111. Clemen CS, et al. Strumpellin is a novel valosin-containing protein binding partner linking hereditary spastic paraplegia to protein aggregation diseases. Brain. 2010;133(10):2920–41.
112. Hirata H, et al. Accordion, a zebrafish behavioral mutant, has a muscle relaxation defect due to a mutation in the ATPase Ca2+ pump SERCA1. Development. 2004;131(21):5457–68.
113. Hirata H, et al. Zebrafish relatively relaxed mutants have a ryanodine receptor defect, show slow swimming and provide a model of multi-minicore disease. Development. 2007;134(15):2771–81.
114. Hirata H, et al. Connexin 39.9 protein is necessary for coordinated activation of slow-twitch muscle and normal behavior in zebrafish. J Biol Chem. 2012;287(2):1080–9.
115. Horstick EJ, et al. Stac3 is a component of the excitation-contraction coupling machinery and mutated in native American myopathy. Nat Commun. 2013;4:1952.
116. Kessler M, et al. A zebrafish model for FHL1-opathy reveals loss-of-function effects of human FHL1 mutations. Neuromuscul Disord. 2018;28(6):521–31.
117. Naganawa Y, Hirata H. Developmental transition of touch response from slow muscle-mediated coilings to fast muscle-mediated burst swimming in zebrafish. Dev Biol. 2011;355(2):194–204.

118. Radev Z, et al. A TALEN-exon skipping design for a Bethlem Myopathy Model in zebrafish. PLoS One. 2015;10(7):e0133986.
119. Ruparelia AA, et al. Characterization and investigation of zebrafish models of filamin-related myofibrillar myopathy. Hum Mol Genet. 2012;21(18):4073–83.
120. Saint-Amant L, et al. The zebrafish ennui behavioral mutation disrupts acetylcholine receptor localization and motor axon stability. Dev Neurobiol. 2008;68(1):45–61.
121. Telfer WR, et al. Neb: a zebrafish model of nemaline myopathy due to nebulin mutation. Dis Model Mech. 2012;5(3):389–96.
122. Zhou W, et al. Identification and expression of voltage-gated calcium channel beta subunits in zebrafish. Dev Dyn. 2008;237(12):3842–52.
123. Zhou W, et al. Non-sense mutations in the dihydropyridine receptor beta1 gene, CACNB1, paralyze zebrafish relaxed mutants. Cell Calcium. 2006;39(3):227–36.
124. Nada SE, Williams FE, Shah ZA. Development of a novel and robust pharmacological model of Okadaic acid-induced Alzheimer's disease in zebrafish. CNS Neurol Disord Drug Targets. 2016;15(1):86–94.
125. Kim YH, et al. Scopolamine-induced learning impairment reversed by physostigmine in zebrafish. Neurosci Res. 2010;67(2):156–61.
126. Abramsson A, et al. The zebrafish amyloid precursor protein-b is required for motor neuron guidance and synapse formation. Dev Biol. 2013;381(2):377–88.
127. Musa A, Lehrach H, Russo VA. Distinct expression patterns of two zebrafish homologues of the human APP gene during embryonic development. Dev Genes Evol. 2001;211(11):563–7.
128. Pu YZ, et al. Generation of Alzheimer's disease transgenic zebrafish expressing human APP mutation under control of zebrafish appb promotor. Curr Alzheimer Res. 2017;14(6):668–79.
129. Anichtchik OV, et al. Neurochemical and behavioural changes in zebrafish *Danio rerio* after systemic administration of 6-hydroxydopamine and 1-methyl-4-phenyl-1,2,3,6-tetrahydropyridine. J Neurochem. 2004;88(2):443–53.
130. Bretaud S, Lee S, Guo S. Sensitivity of zebrafish to environmental toxins implicated in Parkinson's disease. Neurotoxicol Teratol. 2004;26(6):857–64.
131. Lam CS, Korzh V, Strahle U. Zebrafish embryos are susceptible to the dopaminergic neurotoxin MPTP. Eur J Neurosci. 2005;21(6):1758–62.
132. Bretaud S, et al. p53-dependent neuronal cell death in a DJ-1-deficient zebrafish model of Parkinson's disease. J Neurochem. 2007;100(6):1626–35.
133. Edson AJ, et al. Dysregulation in the brain protein profile of zebrafish lacking the Parkinson's disease-related protein DJ-1. Mol Neurobiol. 2019;56(12):8306–22.
134. Hughes GL, et al. Machine learning discriminates a movement disorder in a zebrafish model of Parkinson's disease. Dis Model Mech. 2020;13(10):dmm045815.
135. Merhi R, et al. Loss of Parla function results in inactivity, olfactory impairment, and dopamine neuron loss in zebrafish. Biomedicine. 2021;9(2):205.
136. Milanese C, et al. Hypokinesia and reduced dopamine levels in zebrafish lacking beta- and gamma1-synucleins. J Biol Chem. 2012;287(5):2971–83.
137. Prabhudesai S, et al. LRRK2 knockdown in zebrafish causes developmental defects, neuronal loss, and synuclein aggregation. J Neurosci Res. 2016;94(8):717–35.
138. Sanderson LE, et al. Bi-allelic variants in HOPS complex subunit VPS41 cause cerebellar ataxia and abnormal membrane trafficking. Brain. 2021;144(3):769–80.
139. Van Laar VS, et al. Alpha-Synuclein amplifies cytoplasmic peroxide flux and oxidative stress provoked by mitochondrial inhibitors in CNS dopaminergic neurons in vivo. Redox Biol. 2020;37:101695.
140. Yu T, et al. Behavioral effects of early-life exposure to perfluorooctanoic acid might synthetically link to multiple aspects of dopaminergic neuron development and dopamine functions in zebrafish larvae. Aquat Toxicol. 2021;238:105926.
141. Flinn L, et al. Zebrafish as a new animal model for movement disorders. J Neurochem. 2008;106(5):1991–7.
142. Flinn L, et al. Complex I deficiency and dopaminergic neuronal cell loss in parkin-deficient zebrafish (*Danio rerio*). Brain. 2009;132(Pt 6):1613–23.

143. Flinn LJ, et al. TigarB causes mitochondrial dysfunction and neuronal loss in PINK1 deficiency. Ann Neurol. 2013;74(6):837–47.
144. Soman S, et al. Inhibition of the mitochondrial calcium uniporter rescues dopaminergic neurons in pink1(−/−) zebrafish. Eur J Neurosci. 2017;45(4):528–35.
145. Hirata H, et al. Defective escape behavior in DEAH-box RNA helicase mutants improved by restoring glycine receptor expression. J Neurosci. 2013;33(37):14638–44.
146. Hirata H, et al. Zebrafish bandoneon mutants display behavioral defects due to a mutation in the glycine receptor beta-subunit. Proc Natl Acad Sci U S A. 2005;102(23):8345–50.
147. Nakano Y, et al. Biogenesis of GPI-anchored proteins is essential for surface expression of sodium channels in zebrafish Rohon-beard neurons to respond to mechanosensory stimulation. Development. 2010;137(10):1689–98.
148. Ochenkowska K, Herold A, Samarut E. Zebrafish is a powerful tool for precision medicine approaches to neurological disorders. Front Mol Neurosci. 2022;15:944693.
149. Ogino K, et al. RING finger protein 121 facilitates the degradation and membrane localization of voltage-gated sodium channels. Proc Natl Acad Sci U S A. 2015;112(9):2859–64.

Chapter 2
Unconstrained Biosignal Measurement Technology and Its Applications in Assisting Aged Society

Yosuke Kurihara, Yuri Hamada, and Takashi Kaburagi

Abstract The elderly population in developed countries experiences several issues, such as increased social security costs, the burden of nursing care, and a decrease in the working-age population that can support the elderly. Hence, the government can ensure the safety and security of the elderly by improving various systems; the industry and academia can contribute by providing scientific and technological assistance. Under these circumstances, we have developed biosignal measurement technologies to assist elderly individuals in living healthy lives in an aged society. Our research on biosignal measurement can be divided into two categories: wearable measurement systems that measure biosignal information with high precision by placing sensors on the human body and unconstrained (environmentally installed) measurement systems that measure biosignal information by installing sensors in the environment and not on the human body. In this chapter, we especially introduce unconstrained biosignal measurement technologies that have applications in heart sound detection during sleep and fall detection in elderly individuals.

Keywords Biosignal · Cardiac auscultation · Doppler sensor · Fall detection · Pressure sensor

Y. Kurihara (✉) · Y. Hamada
Department of Industrial and Systems Engineering, College of Science and Technology, Aoyama Gakuin University, Chuo-ku, Sagamihara-shi, Kanagawa, Japan
e-mail: kurihara@ise.aoyama.ac.jp; hamada@ise.aoyama.ac.jp

T. Kaburagi
Department of Natural Sciences, International Christian University, Mitaka-shi, Tokyo, Japan
e-mail: kabutakashi@icu.ac.jp

2.1 Introduction

We investigated the measurement of biosignal information as an application of measurement technology. Our aim was to develop assistive technologies to help elderly individuals live healthy lives in aged society, covering a wide range of research topics. Our research on biosignal measurement can be divided into two categories: wearable measurement systems that measure biosignal information with high precision by placing sensors on the human body [1–11] and unconstrained (environmentally installed) measurement systems that measure biosignal information by installing sensors in the environment and not on the human body [12–39].

Studies on wearable measurement systems include research on swallowing [1, 2], monitoring lung disorders [3], evaluating brain function [4–6], and evaluating daily activities/movements [7–11]. In a study on swallowing, the elevation movement of the larynx during swallowing was measured using optical sensors and high-sensitivity pressure sensors placed on the neck to evaluate swallowing function [1] and classify the viscosity of the swallowed bolus of food [2]. In a study on lung disorders, we developed a spirometry technique by utilizing a vortex whistle whose sound frequency of blowing air was linearly proportional to the flow rate [3]. In a study that investigated brain function, we measured cerebral blood flow using near-infrared spectroscopy (NIRS) installed in the prefrontal regions to determine the type of memory recall process (verbal/nonverbal recall) [4] and evaluated brain activity according to differences in language proficiency [5, 6]. In addition, in research on daily activities/movements, the subject's movements were assessed with accelerometers and gyro sensors installed on the body to estimate the functional independence measure (FIM), an evaluation index for the activities of daily living (ADL) of patients undergoing rehabilitation [7], construct motor learning models during walking [8], analyze the gaits of patients with Parkinson's disease [9, 10], and estimate metabolic equivalents (METs), which is the motor intensity of the ADL measure [11].

Studies using unconstrained sensors include research on the development of unconstrained measurement systems [12–23], evaluation of sleep conditions [24–27], determination of respiratory arrest during sleep [28, 29], estimation of scratching time in patients with atopic dermatitis [30–32], detection of falls in elderly people [33–36], and prediction of the amount of urine accumulated in the bladder [37–39]. To establish unconstrained measurement systems, we developed systems for unconstrained measurements of vibrations (pulse, heart sound, respiration, body movement, and scratching) due to the biological activities of humans lying on bed mats, including highly sensitive pressure sensors [12–18], piezoelectric ceramics [19, 20], thermopiles [21], and flow sensors [22, 23]. To evaluate sleep conditions, we developed signal processing methods such as noise removal and feature extraction to measure unconstrained biosignals to estimate sleep stages [24–27], discriminate respiratory arrest [28, 29], and approximate scratching time [30–32]. For fall detection, we have developed methods to measure human motions by installing microwave Doppler sensors in target environments to automatically discriminate between daily activities and fall events [33–36]. To predict urine accumulation in the bladder, we aim to develop a method to forecast how urine accumulates in the bladder after urination [37, 38] by measuring the absorption spectrum of urine using hyperspectral cameras installed in the environment. Moreover, we are investigating methods to estimate hyperspectral images in several hundred dimensions using three-dimensional RGB images [39] for urine accumulation prediction.

In this work, we introduce research on heart sound detection during sleep [18] and fall detection in elderly individuals by the human body tilt model as applications of unconstrained biosignal measurement technologies.

2.2 Development of a Bed-Based Unconstrained Cardiac Auscultation Method

2.2.1 Background

Cardiac information, such as the commonly used electrocardiogram (ECG), phonocardiogram (PCG), and ballistocardiogram (BCG), is important for diagnosing cardiovascular diseases in medical institutions. However, these methods involve physical contact between physicians and patients, increasing the risk of contact-transmissible infections [40]. To decrease the risk of hospital-acquired infections, cardiac information needs to be measured in an unconstrained manner. Moreover, in home health care, an unconstrained system for measuring cardiac information during sleep is desirable for daily monitoring. Thus, monitoring the cardiac information of a person lying on a bed without any sensors attached to the body is useful for diagnosis and health care.

Therefore, we developed a pneumatic sensing device for unconstrained PCG measurements during sleep. This section introduces the physical model of the sensing device utilized in the proposed unconstrained PCG method, the frequency characteristics of the designed sensing device and comparative results with a photoplethysmography (PPG) sensor.

2.2.2 Proposed Method

2.2.2.1 Physical Model of Vibrations Generated by Heart Sounds Propagated Through a Mattress

In the proposed method, a sensing device is placed under the mattress to measure vibrations generated by heart sounds, as shown in Fig. 2.1. To design an appropriate sensing device based on the frequency characteristics, in this section, we consider how the vibrations generated by the heart sounds of a person lying on a bed propagate through the mattress using a physical model. As shown in Fig. 2.1, when a

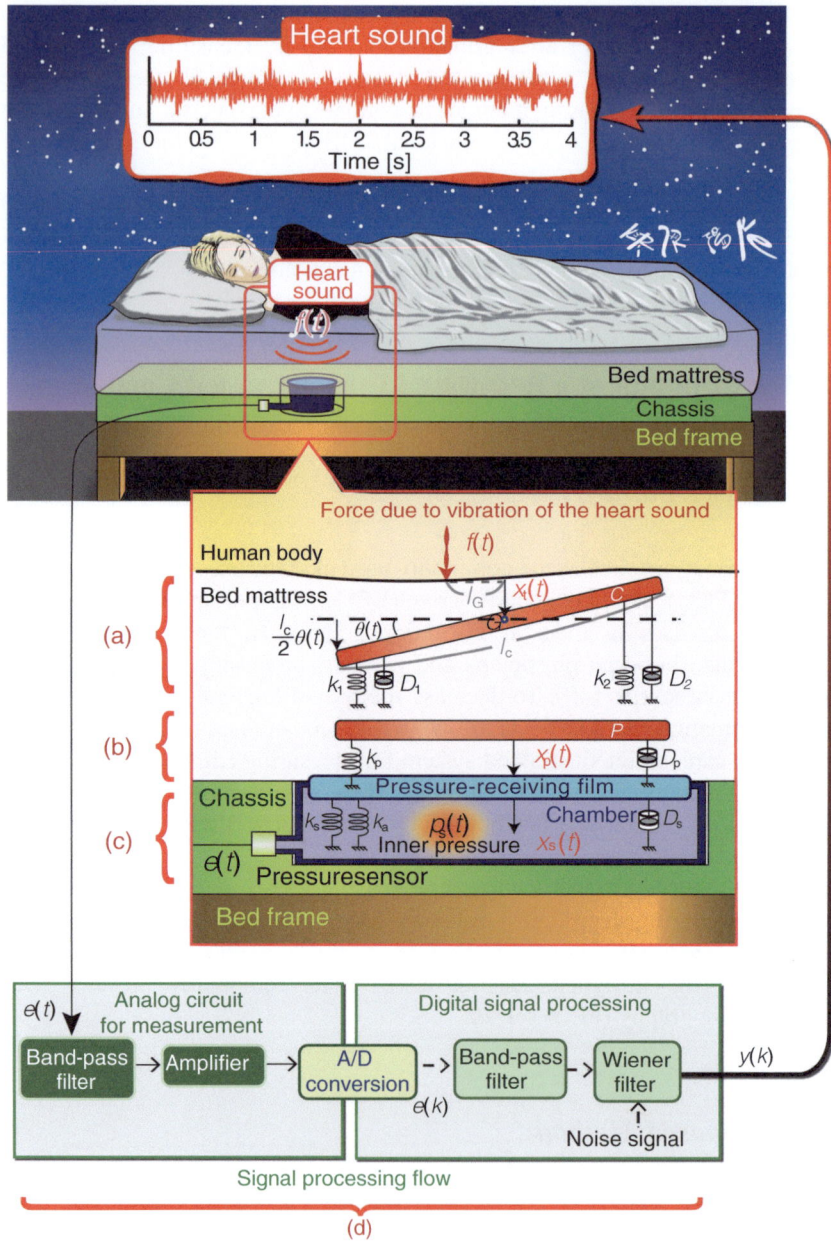

Fig. 2.1 Physical model of the sensing device and the signal processing flow for the proposed unconstrained cardiac auscultation method [18]

person lies on a bed, the vibrations from the heart sounds propagate through the human body and reach the surface of the mattress. Let $f(t)$ be the force of the vibrations generated by the heart sounds acting on the mattress surface at a continuous time t. Figure 2.1a shows a physical model representing how the mattress surface is displaced by $f(t)$. $f(t)$ causes translational motion in the vertical direction and rotational motion around the center of gravity in different areas of the mattress surface. Let C, M_c, G, l_c, l_G, $x_t(t)$, and $\theta(t)$ be the portion of the mattress that moves according to $f(t)$, the mass of C, the center of gravity of C, the length of C, the horizontal displacement from G where $f(t)$ acts vertically, the vertical displacement from G due to translational motion, and the angle between C and the horizontal axis, respectively. Hence, the equation of motion of $x_t(t)$ can be formulated, as shown in Eq. (2.1).

$$M_c \frac{d^2 x_t(t)}{dt^2} = -k_1 x_t(t) - D_1 \frac{dx_t(t)}{dt} - k_2 x_t(t) - D_2 \frac{dx_t(t)}{dt}$$
$$-k_1 \frac{l_c}{2}\theta(t) - D_1 \frac{l_c}{2}\frac{d\theta(t)}{dt} + k_2 \frac{l_c}{2}\theta(t) + D_2 \frac{l_c}{2}\frac{d\theta(t)}{dt} + f(t) \quad (2.1)$$

In Eq. (2.1), k_1 and k_2 represent spring constants, and D_1 and D_2 represent the damping coefficients of C, as shown in Fig. 2.1a. Moreover, the rotational motion of C is shown in (2.2), where I is the moment of inertia.

$$I \frac{d^2 \theta(t)}{dt^2} = -k_1 \left(\frac{l_c}{2}\right)^2 \theta(t) - D_1 \left(\frac{l_c}{2}\right)^2 \frac{d\theta(t)}{dt} - k_2 \left(\frac{l_c}{2}\right)^2 \theta(t)$$
$$-D_2 \left(\frac{l_c}{2}\right)^2 \frac{d\theta(t)}{dt} - k_1 x_t(t) \bullet \frac{l_c}{2} - D_1 \frac{dx_t(t)}{dt} \bullet \frac{l_c}{2} \quad (2.2)$$
$$+k_2 x_t(t) \bullet \frac{l_c}{2} + D_2 \frac{dx_t(t)}{dt} \bullet \frac{l_c}{2} + f(t) \bullet l_G$$

According to Eqs. (2.1) and (2.2), the transfer function $G_1(s)$, where $f(t)$ is the input and $x_t(t)$ is the output, can be calculated as follows:

$$G_1(s) = \frac{\begin{bmatrix} 4Is^2 + \{l_c^2(D_1+D_2) + 2l_c l_G (D_2-D_1)\}s + \\ l_c^2(k_1+k_2) + 2l_c l_G(k_2-k_1) \end{bmatrix}}{\begin{bmatrix} 4M_c Is^4 + (D_1+D_2)(M_c l_c^2+4I)s^3 \\ +\{(M_c l_c^2+4I)(k_1+k_2)+4l_c^2 D_1 D_2\}s^2 \\ +4l_c^2(D_1 k_2+D_2 k_1)s+4l_c^2 k_1 k_2 \end{bmatrix}}. \quad (2.3)$$

The physical model of (2.1) and (2.2) and the transfer function of (2.3) represent the transfer characteristics of the vibrations generated by heart sounds on the mattress near the body. The displacement $x_t(t)$ of C propagates vertically downward through the interior of the mattress as vibrations. Figure 2.1b shows a physical model of the propagation of vibrations due to the displacement $x_t(t)$ to the bottom of the mattress.

Let P, M_p, and $x_p(t)$ be the region inside the mattress where $x_t(t)$ propagates, the mass of P, and the vertical displacement of P at the bottom of the mattress due to the propagation of $x_t(t)$, respectively. The translational motion of $x_p(t)$ is shown in (2.4).

$$M_p \frac{d^2 x_p(t)}{dt^2} = -k_p x_p(t) - D_p \frac{dx_p(t)}{dt} + M_c \frac{d^2 x_t(t)}{dt^2} \quad (2.4)$$

In Eq. (2.4), k_p and D_p represent the spring constant and damping coefficient inside the mattress. Equation (2.4) leads to the transfer function $G_2(s)$, where $x_t(t)$ and $x_p(t)$ are the input and output, respectively, as follows:

$$G_2(s) = \frac{M_c s^2}{M_p s^2 + D_p s + k_p} \quad (2.5)$$

The transfer function $G_B(s)$ of the overall vibration can be formulated as $G_B(s) = G_1(s)G_2(s)$, namely, a system of $G_1(s)$ and $G_2(s)$, with input $f(t)$ and output $x_p(t)$. Because $G_B(s)$ has a low-pass property, with a sixth-order denominator and fourth-order numerator, high-frequency vibrations in $f(t)$ due to heart sounds are attenuated as $x_p(t)$ propagates through the mattress. Therefore, under the mattress, the gain of the heart sounds tends to decrease more due to the low-pass property than the low-frequency pulse wave component.

2.2.2.2 Design of the Sensing Device in the Proposed Unconstrained Cardiac Auscultation System

To improve the decreased gain of the heart sounds under the mattress, we propose a sensing device placed under the mattress, as shown in Fig. 2.1c. The device consists of a pressure film that receives vibrations due to $x_p(t)$ and a chamber sealed by the pressure-receiving film. The chamber is embedded in a chassis that protects the pressure-receiving film by supporting the weight of the human body and mattress. A pressure sensor is installed to measure the internal pressure. The pressure-receiving film is displaced by $x_s(t)$ according to $x_p(t)$. Let M_s, k_s, and D_s be the mass, spring constant, and damping coefficient of the pressure-receiving film and k_a be the spring constant of the air spring in the chamber. The equation of motion for the pressure-receiving film is formulated as follows:

$$M_s \frac{d^2 x_s(t)}{dt^2} = k_p \{x_s(t) - x_p(t)\}$$
$$+ D_p \left\{ \frac{dx_s(t)}{dt} - \frac{dx_p(t)}{dt} \right\} - k_s x_s(t) - D_s \frac{dx_s(t)}{dt} - k_a x_s(t) \quad (2.6)$$

Furthermore, the internal chamber pressure varies with $x_s(t)$. Let $p_s(t)$, P_0, V_0, A_s, and γ be the internal chamber pressure, initial pressure, initial volume, area of the pressure-receiving film, and specific heat ratio, respectively. The following relation between $p_s(t)$ and $x_s(t)$ is obtained:

$$p_s(t) = \frac{k_a}{A_s \gamma} x_s(t) + \left(\frac{V_0 k_a}{\gamma A_s^2} - P_0 \right) \tag{2.7}$$

In Eq. (2.7), the $\left(\frac{V_0 k_a}{\gamma A_s^2} - P_0 \right)$ term is the direct current (DC) component; however, the effect of the DC component is eliminated by a bandpass filter during signal processing. Hence, we ignore the DC component to simplify the model. Equations (2.6) and (2.7) give the transfer function $G_D(s)$, with input $x_p(t)$ and output $p_s(t)$, as follows:

$$G_D(s) = \frac{k_a}{A_s \gamma} \cdot \frac{-(D_p s + k_p)}{M_s s^2 + (D_s - D_p)s + (k_s - k_p + k_a)} \tag{2.8}$$

$$f_n = \frac{1}{2\pi} \sqrt{\frac{k_s - k_p + k_a}{M_s}} \tag{2.9}$$

Hence, the gain in the frequency characteristic in the range of the heart sounds, which is attenuated by $G_B(s)$, can be amplified by an appropriately designed f_n. By measuring $p_s(t)$ with the installed pressure sensor, the heart sounds are the output signal $e(t)$ of the pressure sensor.

2.2.2.3 Signal Processing Flow to Remove Noise

Although the gain is improved by the proposed sensing device, the output signal $e(t)$ measured by the pressure sensor contains both heart sounds and pulsations, the DC component, and white noise caused by the pressure sensor and measurement circuit when installing the bandpass filter. Hence, to obtain the heart sounds, a signal processing flow, as shown in Fig. 2.1d, is applied to $e(t)$. To remove the DC components and prevent aliasing generated through A/D conversion, $e(t)$ is filtered in the range f_{low} to f_{high} by an analog bandpass filter and amplified by a noninverting amplifier. The amplified signal undergoes A/D conversion with sampling frequency f_s. Let $e(k)$ be the discretized signal at time step k, with $e(k)$ containing heart sounds, pulsations, and white noise. To remove pulsations from $e(k)$, a digital bandpass filter with a passband of 20–f_s/2 Hz is applied. Moreover, the filtered signal is passed through a Wiener filter (WF) to remove white noise. Finally, we obtain the heart sound $y(k)$ as the output signal of the WF. To set the optimal variance in the noise signal in the WF algorithm, we calculate the variance according to the signal measured while the patient is unattended on the bed.

2.2.3 Validation Experiment

An experiment was performed to evaluate the accuracy of the proposed method in measuring heart sounds compared with reference heart sounds measured by a stethoscope.

2.2.3.1 Experimental System

In the sensing device, the pressure-receiving film is a diaphragm with a radius of 25.5 mm and a weight of 0.8 g, and f_n is set to approximately 200 Hz. An expanded polystyrene (EPS) board is used as a chassis to support the weights of the human body and mattress. A chamber with the pressure-receiving film is embedded in the EPS, and the diaphragm attached to the chamber is set to protrude slightly horizontally at the EPS height to ensure that the film receives the vibration from the mattress. The chamber's internal pressure is measured by a pressure sensor through an air tube connected to the chamber. The sensing device is placed between the mattress and bed frame, and the participant lies on the mattress. To obtain reference heart sounds and evaluate whether the proposed method accurately measures heart sounds, a stethoscope (3MTM Littmann® Master Cardiology™, 3M Company) was secured to the pulmonic area (the left margin of the second intercostal sternum) of the participant because the pulmonic area can be used to measure both first and second heart sounds. The heart sounds are measured by the stethoscope using the same type of pressure sensor as the one used in the proposed device. The output signals of the proposed device and stethoscope are filtered by analog bandpass filters, with the flow set to 0.008 Hz and f_{high} set to 979 Hz. The filtered signals from the proposed device and stethoscope undergo A/D conversion simultaneously and are stored in a computer. The sampling frequency f_s and measurement time are set to 2 kHz and 20 s, respectively, and $y_r(k)$ is the reference heart sound obtained by the stethoscope.

2.2.3.2 Experimental Procedures

In this experiment, the proposed method was evaluated by comparing $y_r(k)$ for various combinations of bed mattress thicknesses, sleeping postures, and respiratory states. The five mattress thicknesses (T0–T4) were adjusted according to the number of layers in an 8-cm-thick mattress: (T0) 0 cm with no mattress, (T1) 8 cm, (T2) 16 cm, (T3) 24 cm, and (T4) 32 cm. The three sleeping posture conditions (P1–P3) included supine (P1), right-lateral (P2), and left-lateral (P3). The prone position cannot be validated since the stethoscope is attached to the chest. The two respiratory state conditions (R1 and R2) included breathing normally (R1) and stopping breathing (R2). Therefore, the heart sounds were measured for each participant under 30 experimental conditions. The participants included seven males and six females, aged 21 to 59 years (weight: 44—70 kg, height: 152—180 cm). All experiments were performed in accordance with the Life Science Committee of Aoyama Gakuin University (permission no. M15–17). Informed consent was obtained before the experiment began.

2.2.3.3 Evaluation Method

The correlation coefficient between the spectra of $y(k)$ and $y_r(k)$ was used to evaluate accuracy. Moreover, for each experimental condition, the average and standard deviation of the correlation coefficients for all participants were calculated.

2.2.4 Results and Discussion

2.2.4.1 Frequency Response Results

We performed a frequency response experiment to determine whether the theoretical frequency response based on the physical model described in Sect. 2.2.2 was consistent with that of the proposed device and whether the proposed device improves the gain of the heart sounds. In the experiment, three frequency responses, $S_B(f)$, $S_D(f)$, and $S_T(f)$, representing the frequency responses of the vibration propagated through the mattress, the system in which the mattress and sensing device are connected in series, and the theoretical physical model based on connecting the transfer functions $G_B(s)$ and $G_D(s)$ in series, are obtained for comparison. $S_B(f)$ and $S_D(f)$ are obtained by placing a full-range speaker (Fostex FF125WK; Foster Electric Co., Ltd.) that outputs sounds generated by a function generator (AFG-2105, GW Instek) in the frequency range 20 Hz to 1000 Hz on the mattress. The vibrations propagate through the mattress and are measured by a pressure sensor placed under the mattress and the proposed device. Finally, $S_B(f)$ and $S_D(f)$ are the output signals measured by the pressure sensor and the proposed device, respectively. The theoretical frequency response $S_T(f)$ is obtained through a simulation with the transfer functions $G_B(s)$ and $G_D(s)$. The gains of the frequency responses $S_B(f)$, $S_D(f)$, and $S_T(f)$ are normalized between 0 and 1 to perform comparisons in the same range. Figure 2.2a shows the results of the three frequency responses, with the green, red, and blue lines representing $S_T(f)$, $S_D(f)$, and $S_B(f)$, respectively. The resonance frequency appears at 200 Hz in the theoretical frequency response $S_D(f)$ and the frequency response of the proposed device $S_T(f)$. Below 100 Hz, the gain in $S_T(f)$ is larger than that in $S_D(f)$; however, above 200 Hz, the gain in $S_T(f)$ is slightly less than that in $S_D(f)$, although the gain decreases in both responses. Furthermore, in $S_B(f)$, the resonance frequency appears at 50 Hz, and the gain is attenuated in the high-frequency band above 50 Hz, whereas in $S_D(f)$, the resonant frequency f_n appears at 200 Hz, and the gain in the heart sounds is improved by the proposed device.

Fig. 2.2 Results of the validation experiments [18]

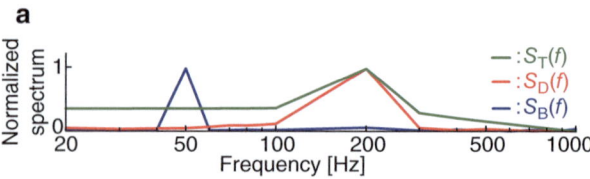

Results of the frequency responses $S_T(f)$, $S_D(f)$, and $S_B(f)$

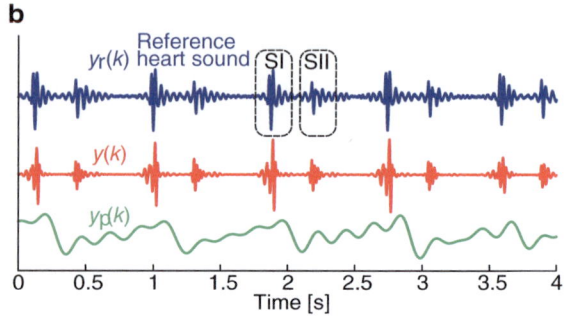

Signal examples of $y(k)$, $y_r(k)$, and $y_p(k)$

2.2.4.2 Heart Sound Measurement Experiment

In addition, to compare our results with those of a PPG sensor, we conducted an additional experiment with two males and two females. The participants wore a PPG sensor on their index finger and a reference stethoscope. Let $y_p(k)$ be the output signal of the PPG sensor. The mattress thicknesses (T0), (T1), and (T2), sleeping postures (P1), (P2), and (P3), and respiratory state (R1) were investigated. Different combinations of (T0–T2) and (P1–P3) were used to establish nine experimental conditions.

The experiment took 45 min for each participant, and the database was 192 MB. Figure 2.2b shows examples of $y(k)$, $y_r(k)$ and $y_p(k)$ for four seconds under the experimental condition T0–P3–R1. In the reference heart sound $y_r(k)$, the first (SI) and second (SII) heart sounds appear periodically; however, in $y(k)$, SI and SII are synchronized. Moreover, the components synchronizing SI and SII in $y(k)$ consistently appear before and after decreases in $y_p(k)$. The remaining 35 data points showed equivalent characteristics, demonstrating the consistency of the proposed method and PPG sensor.

Table 2.1 lists the average and standard deviation of the correlation coefficients for all 30 experimental conditions. The correlation coefficients range from 0.66 to 0.79 and 0.68 to 0.81 in conditions R1 and R2, respectively. Almost all mattress thickness and sleeping posture combinations showed higher correlation coefficients in R2 than R1. Moreover, the average and standard deviation of the correlation coefficients for all 30 experimental conditions were calculated as 0.75 ± 0.12. The proposed method showed no substantial difference in heart sound accuracy for various combinations of mattress thickness, sleeping posture, and respiratory state. Thus, the following four investigations are needed: (1) evaluating other auscultation areas,

Table 2.1 Average and standard deviation of the correlation coefficients for all experimental conditions

		T0	T1	T2	T3	T4
R1	P1	0.66 ± 0.27	0.79 ± 0.07	0.79 ± 0.06	0.74 ± 0.13	0.79 ± 0.07
	P2	0.75 ± 0.13	0.78 ± 0.05	0.77 ± 0.09	0.77 ± 0.06	0.75 ± 0.08
	P3	0.73 ± 0.06	0.76 ± 0.07	0.76 ± 0.05	0.74 ± 0.05	0.76 ± 0.04
R2	P1	0.68 ± 0.25	0.78 ± 0.08	0.79 ± 0.06	0.75 ± 0.16	0.80 ± 0.09
	P2	0.81 ± 0.07	0.81 ± 0.05	0.79 ± 0.06	0.79 ± 0.07	0.76 ± 0.08
	P3	0.78 ± 0.06	0.77 ± 0.06	0.77 ± 0.04	0.73 ± 0.08	0.74 ± 0.07

(2) verifying the effect of background noise in the heart sound frequency range in noisy environments, (3) developing an algorithm for detecting abnormal heart sounds, and (4) analyzing the pathology of the measured heart sounds.

2.3 Unconstrained Fall Detection Method

2.3.1 Background

The risk of falling is a serious issue, especially for elderly individuals that live alone. The WHO suggests that approximately 30% of people over the age of 65 experience falls each year, which is responsible for 40% of injury-related deaths [41]. Therefore, a real-time fall detection method should be developed. Although wearable devices can detect falls with relatively high accuracy, these devices may impose restrictions on daily life, leading to increased mental burdens for caregivers. Therefore, fall detection using environmental devices that do not need to be directly attached to the caregiver's body has been researched. Among such studies, we focus on the microwave Doppler sensor, which is environmentally installed as a sensing device [33]. However, in a previous study on microwave Doppler sensors, non-turning movements were easily misidentified as falls. This problem occurred because the measured data were fundamentally different than data measured by wearable devices. Microwave Doppler sensors can acquire frequencies according to the target speed by applying frequency analysis techniques. However, physical quantities of the body cannot be obtained when the subject falls, which can be acquired by wearable devices. Assuming that microwaves are irradiated to the whole body, the frequency obtained by the microwave Doppler sensor is expected to depend on the speeds of various body parts, such as the arms, legs, and torso. The nature of these data may impact the erroneous identification of non-falling actions as falls.

To address this issue, we wondered if the trunk tilt angle during a fall could be obtained using a microwave Doppler sensor.

Here, we present a method for constructing a model to represent the tilt angle of the body trunk during a fall. We also apply the model to a particle filter to estimate the trunk tilt angle, which cannot be observed directly.

2.3.2 Proposed Method

Figure 2.3 shows the scheme of the proposed method. The method can be divided into three components. In component (A), the human body tilt during movement is modeled based on an inverted pendulum. We focus on the angle as the desired feature. In component (B), the observations of the microwave Doppler sensor are modeled. In component (C), a signal processing flow is applied to discriminate fall/non-fall events based on the proposed model.

2.3.2.1 Human Body Tilt Model

To discriminate fall/non-fall events, we investigate the human body tilt during movement. As shown in Fig. 2.3a, we model the human body tilt based on an inverse pendulum. The body tilt in the continuous time domain t is defined as the angle between the body and the direction of the gravitational acceleration, where $\theta(t)$ is the angle. In the human body tilt model, we consider a state transition equation where $\theta(t)$ is a state variable. Let m, l, I, and g be the mass and height of the participant, the moment of inertia, and gravitational acceleration, respectively. In the model, the kinetic energy $K(t)$ is given by

$$K(t) = \frac{1}{2} I \left\{ \frac{d\theta(t)}{dt} \right\}^2 = \frac{1}{2} m l^2 \left\{ \frac{d\theta(t)}{dt} \right\}^2 \tag{2.10}$$

The potential energy $U(t)$ is given by

$$U(t) = mgl \cos \theta(t) \tag{2.11}$$

Therefore, the Lagrangian $Q(t)$, namely, the difference between the kinetic and potential energies, is

$$Q(t) = \frac{1}{2} m l^2 \left\{ \frac{d\theta(t)}{dt} \right\}^2 - mgl \cos \theta(t) \tag{2.12}$$

According to Eq. (2.12), we obtain the following Lagrange equation:

$$\frac{d^2 \theta(t)}{dt^2} = \frac{g}{l} \sin \theta(t) \tag{2.13}$$

Here, let $\omega(t)$ be $\frac{d\theta(t)}{dt}$. We therefore obtain the following equation as a state transition equation for the human body tilt model:

$$\frac{d}{dt} \begin{bmatrix} \theta(t) \\ \omega(t) \end{bmatrix} = \begin{bmatrix} \omega(t) \\ \frac{g}{l} \sin \theta(t) \end{bmatrix} \tag{2.14}$$

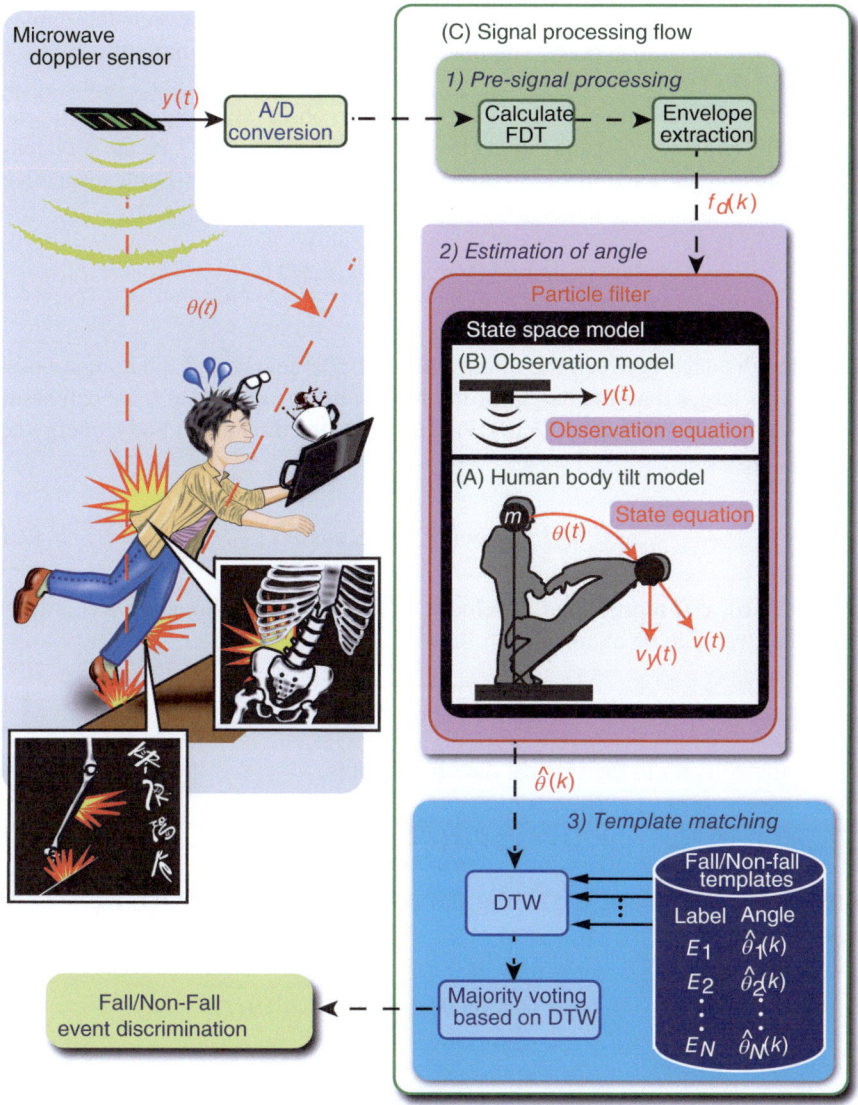

Fig. 2.3 Signal processing flow for the proposed method. The method has three key components: (**a**) Human body tilt model; (**b**) Observation model for the microwave Doppler sensor; (**c**) Signal processing flow for discriminating fall/non-fall events

2.3.2.2 Observation Model for Microwave Doppler Sensor

We apply a microwave Doppler sensor to discriminate fall/non-fall events in the proposed method. Hence, in this section, we develop a model for observing how the human body tilt $\theta(t)$ is measured by the microwave Doppler sensor. As shown in

Fig. 2.3b, the Doppler sensor is attached above the human body at a distance R, and the sensor emits microwaves downward with an angle φ between the emission direction and the human body. The microwave Doppler sensor outputs a signal $y(t)$ containing the Doppler shifted frequency $f_d(t)$, which is proportional to the vertical velocity $v_y(t)$ of the human body. We construct an observation equation that relates the shifted frequency $f_d(t)$ and human body tilt $\theta(t)$. Let λ, f_m and c be the wavelength of the transmitted wave, transmission frequency, and speed of light, respectively. The relation between $f_d(t)$ and $v_y(t)$ is:

$$f_d(t) = \frac{2v_y(t)}{\lambda}\cos\varphi = \frac{2f_m v_y(t)}{c}\cos\varphi \qquad (2.15)$$

When the distance between the microwave Doppler sensor and the human body is sufficiently larger than the height of the human body, i.e., $R >> l$, we can assume that $\phi \approx 0$ ($\cos\phi \approx 1$) in Eq. (2.15). Furthermore, for the human body, energy conservation gives

$$mgl = \frac{1}{2}mv(t)^2 + mgl\cos\theta(t) \qquad (2.16)$$

In Eq. (2.16), $v(t)$ represents the velocity in the tangential direction. According to Eq. (2.16), $v(t)$ can be obtained as follows:

$$v(t) = \sqrt{2gl\{1-\cos\theta(t)\}} \qquad (2.17)$$

Hence, the velocity in the vertical direction $v_y(t)$ is

$$v_y(t) = \sin\theta(t)\sqrt{2gl\{1-\cos\theta(t)\}} \qquad (2.18)$$

According to (2.15) and (2.18), the shifted frequency $f_d(t)$ and $\theta(t)$ are related as follows:

$$f_d(t) = \frac{2f_m}{c}\sin\theta(t)\sqrt{2gl\{1-\cos\theta(t)\}} \qquad (2.19)$$

2.3.2.3 Signal Processing Flow to Discriminate Fall/Non-Fall Events

In this section, we describe the signal processing flow for discriminating fall/non-fall events according to the output signal $y(t)$ of the microwave Doppler sensor based on the state transition equation and observation equation shown in Eqs. (2.14) and (2.19), respectively. As shown in Fig. 2.3c, the signal processing flow has three main components: (1) pre-signal processing; (2) estimation of the human body tilt angle by a particle filter; and (3) template matching to discriminate fall/non-fall events through dynamic time warping.

Pre-Signal Processing

Here, we describe the pre-signal processing approach, which calculates the Doppler shifted frequency $f_d(t)$ according to the output signal $y(t)$ of the microwave Doppler sensor. Let $y(k)$ ($k = 1,2, ..., K$) be the A/D converted signal with the sampling interval Δt. Let k and K be the discrete time step and number of data points. To extract the Doppler shifted frequency $f_d(k)$ in the discrete time domain k, we calculate the FDT [33] from $y(k)$. Since the FDT contains high-frequency noise, an envelope extraction processing is applied, and the envelope signal is $f_d(k)$.

Estimation of the Human Body Tilt Angle with a Particle Filter

To estimate the human body tilt angle $\theta(k)$, we apply a particle filter to the Doppler shifted frequency $f_d(k)$. Let $n_\theta(k)$ and $n_\omega(k)$ be system noise in the state transition equation (Eq. (2.14) and $n_f(k)$ be observation noise in the observation equation (Eq. (2.19)). We assume that the noise $n_\theta(k)$, $n_\omega(k)$ and $n_f(k)$ follow white Gaussian noise distributions with variances σ_θ^2, σ_ω^2 and σ_f^2, respectively. Since Eqs. (2.14) and (2.19) are nonlinear, we apply a filtering method using a particle filter to estimate the true states according to a nonlinear model. Furthermore, since Eq. (2.14) is a nonlinear differential equation in the continuous time domain, the difference equation in the discrete time domain related to the state transition equation is obtained by the fourth–order Runge–Kutta method. Therefore, the state transition equation can be discretized with the sampling interval Δt as follows:

$$\theta(k+1) = \theta(k) + \frac{\Delta t}{6}\{a_1 + 2a_2 + 2a_3 + a_4\} + n_\theta(k) \quad (2.20)$$

$$\omega(k+1) = \omega(k) + \frac{\Delta t}{6}\{b_1 + 2b_2 + 2b_3 + b_4\} + n_\omega(k) \quad (2.21)$$

where

$$a_1 = \omega(k), \quad a_2 = a_1 + \frac{b_1}{2}\Delta t, \quad a_3 = a_1 + \frac{b_2}{2}\Delta t, \quad a_4 = a_1 + b_3\Delta t$$

$$b_1 = \frac{g}{l}\sin\{\theta(k)\}, \quad b_2 = \frac{g}{l}\sin\left\{\theta(k) + \frac{a_1}{2}\Delta t\right\},$$

$$b_3 = \frac{g}{l}\sin\left\{\theta(k) + \frac{a_2}{2}\Delta t\right\}, \quad b_4 = \frac{g}{l}\sin\{\theta(k) + a_3\Delta t\}$$

In addition, the discretized observation equation is obtained as follows:

$$f_d(k) = \frac{2f_m}{c}\sin\theta(k)\sqrt{2gl\{1-\cos\theta(k)\}} + n_f(k) \quad (2.22)$$

Hence, by applying the particle filter with Eqs. (2.21) and (2.22) to the observed Doppler shifted frequency $f_d(k)$, we can estimate the human body tilt angle $\hat{\theta}(k)$.

Template Matching to Discriminate Fall/Non-Fall Events Through Dynamic Time Warping

To discriminate fall/non-fall events, we utilize dynamic time warping as our template matching method to the estimated $\hat{\theta}(k)$. Let $E \in (1, 0)$ be an event, with 1 and 0 representing fall and non-fall events, respectively. As templates, N sets of $\hat{\theta}_i(k)$ ($i = 1, 2, ..., N$) labeled E_i are prepared according to the measurements of fall and non-fall events. When we obtain a new $f_{d,\text{new}}(k)$ with an unknown event E_{new}, $\hat{\theta}_{\text{new}}(k)$ is calculated by the particle filter, and the DTW distance D_i between $\hat{\theta}_{\text{new}}(k)$ and all templates $\hat{\theta}_i(k)$ is calculated as follows

$$D_i := \text{DTW}\left\{\hat{\theta}_{\text{new}}(k), \hat{\theta}_i(k)\right\} \tag{2.23}$$

Even if $\hat{\theta}_i(k)$ and $\hat{\theta}_{\text{new}}(k)$ have the same number of data points K, the timing of fall and non-fall movements differ. The DTW can be performed with different length time-series data, thus permitting time domain drifts between $\hat{\theta}_i(k)$ and $\hat{\theta}_{\text{new}}(k)$. After the DTW distances are obtained, the N_v of E_i corresponding to $\hat{\theta}_i(k)$ in ascending order of DTW distance are selected. Finally, the number of non-fall events and fall events among the N_v events are counted, and the event E_{new} is discriminated to be the event with the higher count.

2.3.3 Validation Experiment

2.3.3.1 Experimental System and Parameter Settings

In this experiment, we utilized a microwave Doppler sensor (InnoSenT: IPS154) to detect human body tilt angles. The sensor was installed at 2.5 m above ground level because the sensor should be incorporated into the ceiling. The sensor records measurements in an approximately 1.5 m radius. The carrier frequency and size of the sensor were 24.125 GHz and 44.0 mm × 30.0 mm × 8.3 mm. The signal obtained by the sensor was stored on a laptop PC as $y(k)$ using an A/D converter (Contec Co., Ltd.: AIO-160802AY-USB). The sampling interval Δt was set to 1/40,000 s. The WT parameters were set as follows: $f_{\text{low}} = 20$ Hz; $f_{\text{high}} = 960$ Hz; and $\Delta f = 20$ Hz. The mother wavelet in the WT was the Meyer function. The PF parameters were set as follows: number of particles $p_n = 200$; variance in system noise in angle $\sigma_\theta^2 = 10^{-3}$; variance in system noise in angular velocity $\sigma_\omega^2 = 10^{-3}$; variance in observation noise $\sigma_f^2 = 10^{-1}$; number of selected DTW $N_v = 17$; inverted pendulum height $l = 1.7$ m;

and inverted pendulum weight $m = 65$ kg. The choice criterion of each particle was the weighted mean. The subjects in the experiment included 20 males who provided informed consent. The subjects' ages and heights ranged from 20 to 24 years and 1.63 m to 1.80 m.

2.3.3.2 Measurement Procedure

In this experiment, we measured three types of fall movements: (A1) tripping, (A2) slipping, and (A3) fainting. Eight trials were performed for each fall movement, yielding 24 data points per person. To measure each fall movement, we used a mattress as a physical injury-prevention measure. We also measured four types of non-fall movements: (B1) walking, (B2) bending, (B3) sitting, and (B4) standing. Six trials were performed for each non-fall movement, yielding 24 data points per person. The measurement time was 10 s ($T_m = 10$, $K = T_m/\Delta t = 40{,}000$). The fall event data included 480 data points (20 subjects × 3 types of movement × 8 trials) and were labeled "1." The non-fall event data included 480 data points (20 subjects × 4 types of movement × 6 times) and were labeled "0."

2.3.3.3 Evaluation

We utilized dynamic time warping (DTW) as our template matching method to discriminate fall/non-fall events. This method was applied to $\hat{\theta}(k)$, which was obtained with the proposed method. To evaluate the proposed method, we implemented leave-one-out cross-validation in the DTW. To apply this method, we divided the template data into two datasets: testing data and training data. First, we arbitrarily extracted one $\hat{\theta}(k)$ from the template data. This $\hat{\theta}(k)$ was regarded as the testing set, and the remaining data were regarded as the training set. This process was repeated 960 times, and all data were defined as testing data at least one time. Then, we labeled each testing data as "fall" or "non-fall" according to the DTW results. Finally, we discriminated fall and non-fall events based on the obtained labels. In this research, we calculated four metrics: true positive (TP), false positive (FP), true negative (TN), and false negative (FN). Then, we calculated the *accuracy*, positive predictive value (PPV), negative predictive value (NPV), *sensitivity, specificity* and *F* value (*F*), which are defined below:

$$
\begin{aligned}
\text{Accuracy} &= (TP+TN)/(TP+FP+TN+FN) \\
\text{PPV} &= (TP)/(TP+FP) \\
\text{NPV} &= (TN)/(TN+FN) \\
\text{Sensitivity} &= (TP)/(TP+FN) \\
\text{Specificity} &= (TN)/(TN+FP) \\
F &= (2TP)/(2TP+FP+FN)
\end{aligned}
\quad (2.24)
$$

Table 2.2 Confusion matrix for the proposed method

		True state	
		Fall	Non-fall
Estimated state	Fall	TP = 473	FP = 24
	Non-fall	FN = 7	TN = 456

In this study, 20 subjects were included. We evaluated the proposed method by calculating the average value of all subjects.

2.3.4 Results

Table 2.2 shows the confusion matrix of the proposed method. According to this table, we obtain ACC = 0.97, PPV = 0.95, NPV = 0.99, Sens = 0.99, Spec = 0.95, and F = 0.97.

2.4 Conclusion

In this work, we briefly introduce wearable and unconstrained biosignal measurement techniques. We also discuss heart sound detection during sleep and fall detection in elderly individuals as applications of our proposed unconstrained biosignal measurement system. In recent years, our work and other unconstrained biosignal measurement technologies have been developed with the same accuracy as wearable devices. These systems are expected to enable stress-free monitoring of biosignal information in elderly individuals. Aging/aged individuals in developed countries face various issues, including increased social security costs, increased nursing care burdens, and decreases in the working-age population that supports elderly individuals. To improve the safety and security of elderly individuals, it is essential not only for the government to improve various systems but also for industry and academia to provide scientific and technological support. We believe that the research proposed in this work will contribute significantly to solving the problems of aging/aged societies.

Acknowledgments Parts of this work was supported by JSPS KAKENHI Grant Numbers 23760372, 16K16392, 19K20386, and 20K12769.

References

1. Kurihara Y, Watanabe K, Yang Y, Tanaka H. Construction of age model for the evaluation of swallowing function using photo sensors. IEEE Trans Neural Syst Rehabil Eng. 2010;18:515–22. https://doi.org/10.1109/tnsre.2010.2047268.

2. Kurihara Y, Kaburagi T, Kumagai S, Matsumoto T. Development of swallowing-movement-sensing device and swallowing-state-estimation system. IEEE Sensors J. 2019;19:3532–42. https://doi.org/10.1109/jsen.2019.2894744.
3. Yoshida T, Hamada Y, Nakamura S, Kurihara Y, Watanabe K. Spirometer based on vortex whistle to monitor lung disorders. IEEE Sensors J. 2022;22:11162–72. https://doi.org/10.1109/jsen.2022.3170314.
4. Kurihara Y, Kaburagi T, Nishio K, Hamada Y, Matsumoto T, Kumagai S. Discrimination of verbal/visuospatial memory retrieval processes by measuring prefrontal lobe blood volume with functional near-infrared spectrometry. IEEE Access. 2020;8:208683–95. https://doi.org/10.1109/access.2020.3038553.
5. Watanabe K, Niimura Y, Kurihara Y, Watanabe K. Temporal changes in NIRS outputs in prefrontal regions when listening to languages. Artif Life Robot. 2015;20:183–9. https://doi.org/10.1007/s10015-015-0214-0.
6. Watanabe K, Tanaka H, Takahashi K, Niimura Y, Watanabe K, Kurihara Y. NIRS-based language learning BCI system. IEEE Sensors J. 2016;16:2726–34. https://doi.org/10.1109/jsen.2016.2519886.
7. Hirota T, Hamada Y, Kaburagi T, Kurihara Y. Estimation of functional independence measure motor score based on a trunk control model. Int J Affect Eng. 2022;21:101–10. https://doi.org/10.5057/ijae.ijae-d-21-00013.
8. Watanabe K, Tanaka H, Kurihara Y, Watanabe K. A progression and retrogression mathematical model for the motor learning process. IEEE Trans Hum Mach Syst. 2016;46:159–64. https://doi.org/10.1109/thms.2015.2469150.
9. Yoneyama M, Kurihara Y, Watanabe K, Mitoma H. Accelerometry-based gait analysis and its application to Parkinson's disease assessment—part 1: detection of stride event. IEEE Trans Neural Syst Rehabil Eng. 2014;22:613–22. https://doi.org/10.1109/tnsre.2013.2260561.
10. Yoneyama M, Kurihara Y, Watanabe K, Mitoma H. Accelerometry-based gait analysis and its application to parkinson's disease assessment—part 2: a new measure for quantifying walking behavior. IEEE Trans Neural Syst Rehabil Eng. 2013;21:999–1005. https://doi.org/10.1109/tnsre.2013.2268251.
11. Kurihara Y, Watanabe K, Yoneyama M. Estimation of walking exercise intensity using 3-D acceleration sensor. IEEE Trans Syst Man Cybern C Appl Rev. 2012;42:495–500. https://doi.org/10.1109/tsmcc.2011.2130522.
12. Watanabe K, Kurihara Y, Tanaka H. Ubiquitous health monitoring at home—sensing of human biosignals on flooring, on tatami mat, in the bathtub, and in the lavatory. IEEE Sensors J. 2009;9:1847–55. https://doi.org/10.1109/jsen.2009.2030987.
13. Watanabe K, Kurihara Y, Nakamura T, Tanaka H. Design of a low-frequency microphone for mobile phones and its application to ubiquitous medical and healthcare monitoring. IEEE Sensors J. 2010;10:934–41. https://doi.org/10.1109/jsen.2009.2038230.
14. Nakamura T, Kurihara Y, Watanabe K, Terada M. Design of a preamplifier for capacitive sensors with wide low-frequency range and low drift noise. IEEE Sensors J. 2012;12:378–83. https://doi.org/10.1109/jsen.2011.2161282.
15. Kurihara Y, Watanabe K. Suppression of artifacts in biomeasurement system by pneumatic filtering. IEEE Sensors J. 2012;12:416–22. https://doi.org/10.1109/jsen.2010.2073700.
16. Watanabe K, Kurihara Y, Watanabe K, Azami T, Nukaya S, Tanaka H. Biosignals sensing by novel use of bidirectional microphones in a mobile phone for ubiquitous healthcare monitoring. IEEE Trans Hum Mach Syst. 2014;44:545–50. https://doi.org/10.1109/thms.2014.2320945.
17. Kurihara Y, Kaburagi T, Watanabe K. Development of an omnidirectional microphone with high gain in low-frequency band based on a bidirectional microphone. IEEE Sensors J. 2016;16:1772–8. https://doi.org/10.1109/jsen.2015.2504544.
18. Nishio K, Kaburagi T, Hamada Y, Kurihara Y. Development of a bed-based unconstrained cardiac auscultation method. IEEE Sens Lett. 2021;5:1–4. https://doi.org/10.1109/lsens.2021.3096116.
19. Nukaya S, Shino T, Kurihara Y, Watanabe K, Tanaka H. Noninvasive bed sensing of human biosignals via piezoceramic devices sandwiched between the floor and bed. IEEE Sensors J. 2012;12:431–8. https://doi.org/10.1109/jsen.2010.2091681.

20. Nukaya S, Sugie M, Kurihara Y, Hiroyasu T, Watanabe K, Tanaka H. A noninvasive heartbeat, respiration, and body movement monitoring system for neonates. Artif Life Robot. 2014;19:414–9. https://doi.org/10.1007/s10015-014-0179-4.
21. Kurihara Y, Watanabe K. Development of unconstrained heartbeat and respiration measurement system with pneumatic flow. IEEE Trans Biomed Circuits Syst. 2012;6:596–604. https://doi.org/10.1109/tbcas.2012.2189007.
22. Watanabe K, Kurihara Y, Kobayashi K, Suzuki K. In-bed biosignal acquisition from conventional differential pressure sensor based on thermal flow principle. IEEE Sensors J. 2021;21:5340–8. https://doi.org/10.1109/jsen.2020.3031066.
23. Watanabe K, Kurihara Y, Kobayashi K, Suzuki K. Ballistocardiogram (BCG) measurement by a differential pressure sensor. IEEE Sensors J. 2021;21:8583–92. https://doi.org/10.1109/jsen.2020.3046724.
24. Kurihara Y, Watanabe K, Tanaka H. Sleep-states-transition model by body movement and estimation of sleep-stage-appearance probabilities by Kalman filter. IEEE Trans Inf Technol Biomed. 2010;14:1428–35. https://doi.org/10.1109/titb.2010.2067221.
25. Kurihara Y, Watanabe K, Nakamura T, Tanaka H. Unconstrained estimation method of delta-wave percentage included in EEG of sleeping subjects. IEEE Trans Biomed Eng. 2011;58:607–15. https://doi.org/10.1109/tbme.2010.2096559.
26. Kurihara Y, Watanabe K. Sleep-stage decision algorithm by using heartbeat and body-movement signals. IEEE Trans Syst Man Cybern A Syst Hum. 2012;42:1450–9. https://doi.org/10.1109/tsmca.2012.2192264.
27. Kurihara Y, Watanabe K, Tanaka H. A sleep model and an observer using the Lotka–Volsterra equation for real-time estimation of sleep. Artif Life Robot. 2016;21:132–9. https://doi.org/10.1007/s10015-016-0264-y.
28. Kurihara Y, Watanabe K, Kobayashi K, Tanaka H. Unconstrained respiration measurement and respiratory arrest detection by dynamic threshold in transferring patients by stretcher. Electron Commun Jpn. 2011;94:41–50. https://doi.org/10.1002/ecj.10295.
29. Kohama M, Hamada Y, Kaburagi T, Kurihara Y. Classification of unconstrained respiratory states utilising multidimensional probability distribution based on respiratory frequency information at each time step. Int J Affect Eng. 2022;21:93–100. https://doi.org/10.5057/ijae.ijae-d-21-00012.
30. Kurihara Y, Kaburagi T, Watanabe K. Development of a non-contact sensing method for scratching activity measurement. IEEE Sensors J. 2013;13:3325–30. https://doi.org/10.1109/jsen.2013.2264283.
31. Kurihara Y, Kaburagi T, Watanabe K. Sensing method of patient's body movement without attaching sensors on the patient's body -evaluation of "scratching cheek", "turning over and scratching back" and "scratching shin". IEEE Sensors J. 2016;16:8271–8. https://doi.org/10.1109/jsen.2016.2555932.
32. Kaburagi T, Kurihara Y. Algorithm for estimation of scratching time. IEEE Sensors J. 2017;17:2198–204. https://doi.org/10.1109/jsen.2017.2658949.
33. Shiba K, Kaburagi T, Kurihara Y. Fall detection utilizing frequency distribution trajectory by microwave Doppler sensor. IEEE Sensors J. 2017;17:7561–8. https://doi.org/10.1109/jsen.2017.2760911.
34. Shiba K, Kaburagi T, Kurihara Y. Monitoring system to detect fall/non-fall event utilizing frequency feature from a microwave Doppler sensor: validation of relationship between the number of template datasets and classification performance. Artif Life Robot. 2017;23:152–9. https://doi.org/10.1007/s10015-017-0409-7.
35. Kaburagi T, Shiba K, Kumagai S, Matsumoto T, Kurihara Y. Real-time fall detection using microwave Doppler sensor—computational cost reduction method based on genetic algorithm. IEEE Sens Lett. 2019;3:6000404. https://doi.org/10.1109/lsens.2019.2892006.
36. Nishio K, Kaburagi T, Hamada Y, Matsumoto T, Kumagai S, Kurihara Y. Construction of an aggregated fall detection model utilizing a microwave Doppler sensor. IEEE Internet Things J. 2022;9:2044–55. https://doi.org/10.1109/jiot.2021.3089520.

37. Kurihara Y, Yamasaki T, Kaburagi T, Kumagai S, Matsumoto T. Model of urine accumulation in the bladder and method for predicting unconstrained urine volume based on absorption spectrum of urine. IEEE Access. 2020;8:69368–77. https://doi.org/10.1109/access.2020.2986584.
38. Taku H, Yuri H, Takashi K, Yosuke K. Predicting the bladder urinary volume with a reabsorbed primitive urine model. SICE J Control Meas Syst Integr. 2021;14:2–9. https://doi.org/10.1080/18824889.2021.1874679.
39. Sato R, Hamada Y, Kaburagi T, Kurihara Y. Evaluation of colour space effect on estimation accuracy of hyperspectral image by dimension extension based on RGB image. SICE J Control Meas Syst Integr. 2022;15:86–95. https://doi.org/10.1080/18824889.2022.2048532.
40. Jiang L, Ng IHL, Ya H, Li D, Tan LWL, Ho HJA, Chen MIC. Infectious disease transmission: survey of contacts between hospital-based healthcare workers and working adults from the general population. J Hosp Infect. 2018;98:404–11. https://doi.org/10.1016/j.jhin.2017.10.020.
41. WHO Maternal, Newborn, Child and Adolescent Health, and Ageing. WHO global report on falls prevention in older age. Geneva: World Health Organization; 2008.

Chapter 3
Effects of Aging on the Vestibulo-Cardiovascular Reflex

Kunihiko Tanaka

Abstract Gravity for the human body has a major effect on the cardiovascular system. During the postural change from a supine to a standing position, gravity generates a hydrostatic pressure gradient according to the longitudinal direction, decreasing venous return and decreasing cardiac output. Thus, arterial pressure (AP) should be decreased, but healthy individuals usually maintain AP. For this AP control, baroreflex control is considered to be important. However, this control is a negative feedback system and not a maintenance system. The importance of the vestibular system, which is the sensor of acceleration force or gravity, has been investigated to maintain AP or raise AP in advance at the onset of postural change. With the change in gravity, sympathetic nerve activity increases, and AP similarly increases regardless of the direction of gravity. At the onset of head-up tilt in human subjects, vestibular systems, especially otolith organs, sense the changes in the direction of gravity for the body and raise AP in advance reflexively. The raise is balanced with a decrease in AP due to the footward fluid shift, and AP is maintained. A type of orthostatic hypotension that AP decreases at the onset of postural change might be induced by an imbalance or dysfunction of the vestibulo-cardiovascular reflex. In the aged subjects and astronauts after long-term space flight, the control function might be diminished.

Keywords Gravity · Vestibulo-cardiovascular reflex · Stochastic resonance · Galvanic vestibular stimulation · Orthostatic hypotension · Baroreflex

---

K. Tanaka (✉)
Graduate School of Health and Medicine, Gifu University of Medical Science, Gifu, Japan
e-mail: ktanaka@u-gifu-ms.ac.jp

© The Author(s), under exclusive license to Springer Nature Singapore Pte Ltd. 2024, corrected publication 2024
T. Shiozawa et al. (eds.), *Gerontology as an Interdisciplinary Science*, Current Topics in Environmental Health and Preventive Medicine,
https://doi.org/10.1007/978-981-97-2712-4_3

3.1 Introduction

On the earth, human beings live in a 1 G environment. Unlike other quadrupeds, we live with sitting and standing positions in daily life. During the posture, gravity is loaded in the longitudinal direction of the body and has a major effect on the cardiovascular system [1]. During recumbent posture, the body fluid, especially blood, is distributed uniformly along the body, and arterial pressure (AP) is uniform from the head to the foot since the gravitational force for the longitudinal direction is almost zero. However, with a postural change to sitting or standing, the direction of gravity force for the body is changed and generates a hydrostatic pressure gradient according to the body. Hydrostatic pressure on the head is lower, and the pressure on the feet is higher. Thus, arterial and venous pressure on the feet is higher than in the head. With this higher venous pressure, veins in the lower body are dilated since the vessel walls of the veins are highly compliant or very soft. With this dilation, blood is pooled in the lower body, and venous return or cardiac output is decreased. As a result, AP should be decreased since AP is a product of cardiac output and total peripheral resistance of the blood vessels. For maintaining AP during the postural change or standing, baroreflex control is considered important [2, 3]. According to the blood pooling or footward blood shift, the stretch of the baroreceptors in the carotid sinus, aortic arch [4, 5], and those for the low-pressure circulation [6–8] are reduced, resulting in a decrease in discharge. The decreased discharge is treated with the central nervous system and reflexively increases sympathetic nerve activity. The increase raises the cardiac contractility and total peripheral resistance of the vessels, and AP is maintained (Fig. 3.1). However, this reflex control is the negative feedback system, and AP should be decreased to activate the system. In short, the control does not have a preventive effect against the decrease in AP. However, the AP of healthy people is not changed and maintained during postural change and standing [9]. To maintain AP or to raise AP in advance at the onset of postural change, another system is needed. The system senses the postural change and should change AP regardless of the AP change. In the human body, the equilibrium organs sense changes in posture. Equilibrium organs include somatosensory systems such as muscle and tendon spindles, visual sensation, and the vestibular system [10–13].

Among these three systems, an important role of the vestibular system in regulating AP has been reported in animal studies [14–16]. The vestibular system is located in the inner ear and consists of otoliths and the semicircular canals. The otolith organs sense linear acceleration, including gravity, and determine posture [17]. AP is raised with exposure to either hypergravity or microgravity, but the responses were diminished in the animals with the vestibular lesion [14, 15]. Those responses were demonstrated in the rats or four-footed animals with recumbent postures. Thus, the vestibular system raises AP regardless of the direction of gravity of the body. In this chapter, AP control in human subjects via the vestibular system is described.

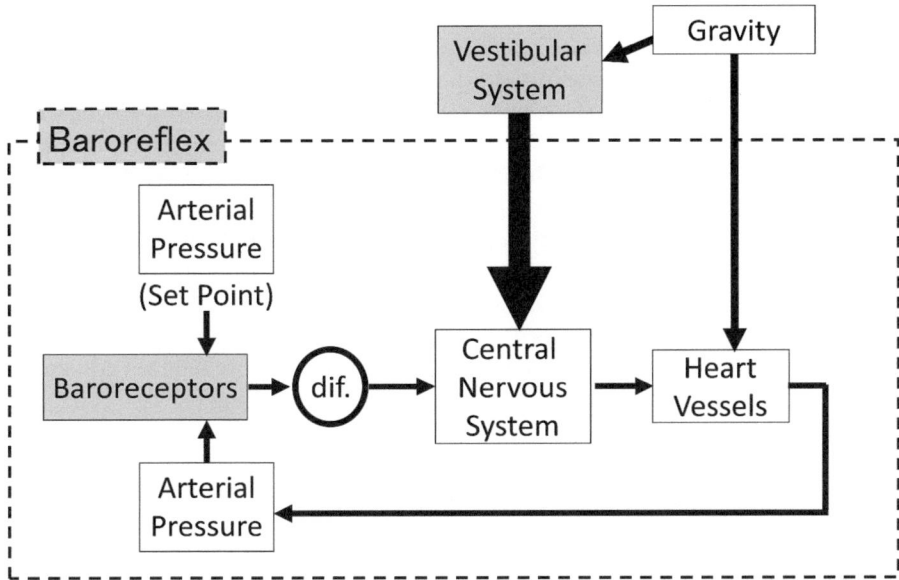

Fig. 3.1 A block diagram of baroreflex control for arterial pressure with gravity change. The baroreceptors sense the difference between the set point of arterial pressure and current arterial pressure. The difference is sent to the central nervous system, and the system changes the dynamics of the cardiovascular system, such as the heart and the blood vessels. The modified arterial pressure is again sensed by baroreceptors. Gravity affects the heart and the vessels via changes in hydrostatic pressure. The sensor of gravity, the vestibular system, also affects the cardiovascular system

3.2 Sympathetic Nerve Activity with Gravity Change

Head-down neck flexion (HDNF) is used for human subjects to stimulate the vestibular system without fluid shift. Subjects lay in a prone position with the neck extended and chin supported. The chin support is removed and the head is lowered to the HDNF position. With HDNF, the direction of gravity for the vestibular system is changed, and muscle sympathetic nerve activity is increased [18–20]. The increase is augmented with orthostatic challenge or fluid shift and baroreceptors unloading using lower body negative pressure (LBNP) [20]. LBNP with −10 and −30 mmHg increase muscle sympathetic nerve activity from 1.5 to 2.5 times, respectively, due to the fluid shift. HDNF during LBNP increased the activity to 2 and 3 times, respectively, compared to baseline. Thus, the vestibule-sympathetic reflex and baroreflex have an additive effect on controlling sympathetic nerve activity. This additive effect should play an important role in controlling sympathetic nerve activity and AP during postural change.

To investigate the effects of changes in gravity on sympathetic nerve activity and AP in human subjects, a parabolic flight is also used. The flight changes gravity in the airplane from 1 G to 2 G at first, and the 2 G condition is obtained for 30 s. Gravity directly reduces from 2 G to 0 G, and the 0 G condition continues for 20 s.

Then, hypergravity of 1.8 G is followed to recover the flight condition [21, 22]. During the flight, at the onset of 2 G, the blood of the sitting subjects is considered to be shifted footward, and venous return and cardiac output should be decreased, but AP and sympathetic nerve activity are significantly increased. At the onset of 0 G, AP further increased, but after that, it gradually decreased. Combined with HDNF studies described above, the increases are derived by vestibular input at the onset of 2 G and 0 G regardless of fluid shift. The footward fluid shift augments the increase in sympathetic nerve activity, and the activity leads to a significant increase in AP. After that, the baroreflex is activated to recover the overly increased AP, and the sympathetic nerve activity is shut down. According to the shutdown, AP decreased overly during 0G regardless of the conceivable headward blood shift. Sympathetic nerve activity is again increased to recover overly reduced AP and to respond to another vestibular input and baroreceptors unloading due to a change in gravity from 0 G to 1.8 G.

3.3 Arterial Pressure Control During Postural Change

During postural change, the direction of gravity for the body is changed as described above. To investigate the role of the vestibular organs in animal studies, vestibular damage model is used, and the responses between intact and damaged models were compared [14, 15, 23]. For human studies, an alternative method for acutely interrupting the vestibular-mediated AP response to gravitational change is required to examine the role of the vestibular system in controlling AP during postural or gravitational change.

Since galvanic vestibular stimulation (GVS) creates an imbalance in the vestibular inputs as well as direction-specific deviation, it has been used for the functional exploration of the vestibular system in animals and humans [24–27]. GVS is an application of the precisely controlled change in the activity of vestibular afferents from all vestibular organs without any potentially confounding changes in non-vestibular inputs [28]. To investigate the role of the vestibular system during HUT, AP is continuously measured during HUT using GVS to interrupt the vestibular inputs due to postural change. Young volunteers were subjected to 60° of HUT and lower body negative pressure LBNP to investigate AP. LBNP induces a footward fluid shift without vestibular stimulation. LBNP of $-55 \sim -60$ mmHg produced a change in the calf circumference, an index of fluid shift similar to that induced by the HUT. On the other hand, the HUT elicits not only a footward fluid shift but also vestibular stimulation on change of the direction of gravity. During HUT, footward fluid shift might decrease the AP due to decreases in the venous return and cardiac output [29]. However, if GVS is not applied, the mean AP is well-maintained during posture transition in young subjects. This observation is consistent with that of another study [9]. On the other hand, HUT with GVS induces a transient decrease in AP of 17 mmHg from that during the supine position at the onset of HUT (Fig. 3.2). The different AP responses to HUT between the absence and presence of

3 Effects of Aging on the Vestibulo-Cardiovascular Reflex

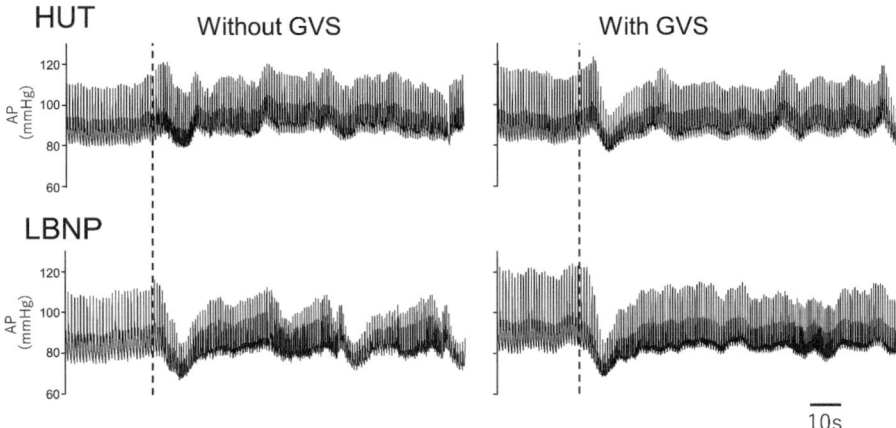

Fig. 3.2 Arterial pressure (AP) during 60° of head-up tilt (HUT; upper panels) and lower body negative pressure (LBNP; lower panels) in a healthy young subject without galvanic vestibular stimulation (GVS; left panel), and with GVS (right panels). Dotted lines are the start of tilt and negative pressure. Without GVS, AP fluctuates but is maintained during HUT. However, AP is decreased at the onset of HUT when GVS is applied. During LBNP, AP is decreased at the onset of negative pressure and gradually recovered regardless with and without GVS

GVS are probably due to the involvement of the vestibular-mediated AP control since GVS attenuates gravitational change-induced AP response, similar to that observed in lesioned vestibular systems [30]. This notion is supported by the observation that a similar decrease in AP is observed at the onset of LBNP, wherein the same amount of footward fluid shift is induced without gravitational change. Further, the AP response to LBNP is not affected by GVS. These results indicate that GVS significantly affects the AP response only if the vestibular system is involved. In this way, GVS can be used as a tool for interrupting vestibular-mediated cardiovascular response in human subjects, vestibular system is also working at the onset of postural change in human subjects.

During HUT, both semicircular canals and otolith organs of the vestibular system are stimulated by pitch rotation and changes in the direction of gravity for the body, respectively [31]. However, it has been considered that the otolith organs are more important for sympathetic nerve activity and AP control during postural change rather than semicircular canals [32, 33]. Indeed, the caloric weakness or deviation of the semicircular canals and the related functions are not correlated with changes in AP upon HUT. However, deviation of the otolith and the related functions examined by subjective visual vertical (SVV) testing correlates well with changes in AP [34]. Thus, the deviation of the otolith and the related functions are proportional to the AP change during HUT. The otolith organs might not correctly sense the change in the direction of gravity for the body due to the deviation, and AP might not be maintained. If the function does not deviate, AP is well maintained during HUT.

After the decrease in AP at the onset of HUT, AP gradually recovers to the control level. The recovery of AP during the late phase of HUT with GVS is considered

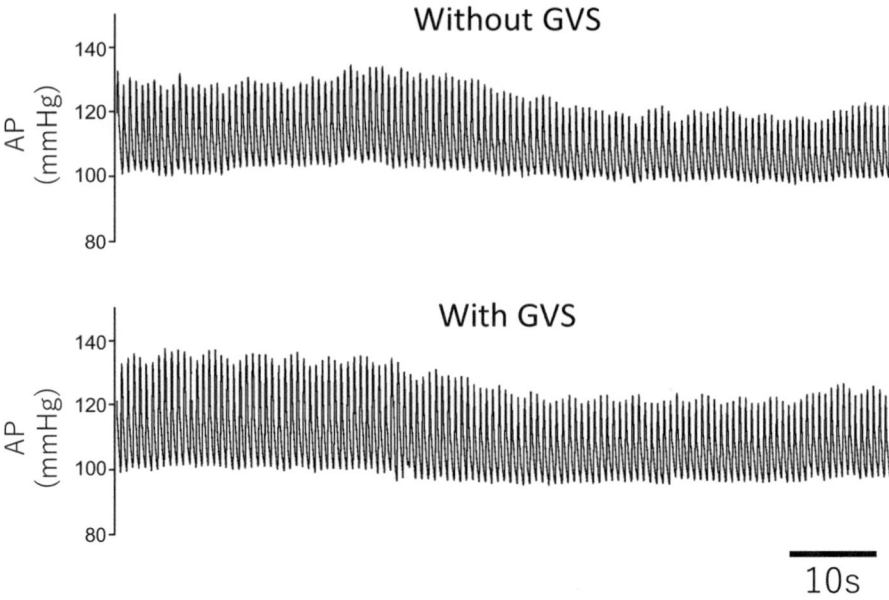

Fig. 3.3 Arterial pressure (AP) during 60° of head-up tilt (HUT) in an aged subject without galvanic vestibular stimulation (GVS) (upper panel), with GVS (lower panel). AP is decreased at the onset of HUT and did not recover during HUT regardless with and without GVS

to be mediated by the baroreflex. The vestibular system is not involved in this period since no difference is detected between HUT with and without GVS [35]. In aged subjects, AP decreases at the onset of HUT regardless of the existence of GVS and does not recover during the measurement (Fig. 3.3). The decrease in AP upon posture transition is probably due to footward fluid shift, and AP maintenance via the vestibular system does not work sufficiently, unlike young subjects. Ray et al. [36] showed that increased muscle sympathetic nerve activity in response to head-down rotation was attenuated in aged subjects compared to younger subjects. Since aging is associated with degenerative changes in vestibular epithelia [37, 38], sensitivity of the vestibular organ might be reduced.

To interrupt the vestibular-mediated cardiovascular response, high-amplitude GVS of 0.5 and 2.0 mA are employed. At these amplitudes, human subjects feel painless head sway [35]. In contrast, some studies have shown that adding small amplitude noise or subsensory electrical stimulation paradoxically improves the signal detection of receptors. Electric stimulation of the auditory nerve improves sound recognition of the auditory brainstem [39]. Subsensory electrical stimulation to the knee joint improves the proprioception of the mechanoreceptors [40]. As a result of this improvement, subsensory stimulation to the ankle muscles and ligaments reflexively improve postural stability [41]. The mechanism is unclear, but the phenomenon that noise increases the quality of detection performance or signal transmission is known as stochastic resonance. According to the concept, when GVS of 0.1 mA below the somatosensory threshold was applied, improved or augmented AP response at the onset of HUT was observed [34]. At the onset of HUT in healthy subjects, mean AP decreased 5 mmHg [34, 35, 42]. Thus, to observe the cardiovascular responses at the onset of HUT, dividing subjects into two groups made the results clearer than analyzing all subjects together. One comprised the subjects whose AP decreased by more than 5 mmHg at the onset of HUT (DOWN), and the other comprised the subjects whose AP increased or decreased by less than 5 mmHg at the onset of HUT (UP). GVS itself does not change, which means AP is in the supine position. Mean AP decreased 12 mmHg from that in the supine position without GVS at the onset of HUT in the DOWN group. Applying GVS at an amplitude of 0.1 mA below the somatosensory threshold (0.3 to 0.7 mA), the decrease in AP was not observed. Thus, subsensory GVS improved AP control or raised AP actively at the onset of HUT, regardless of the degree of footward blood shift (Fig. 3.4).

In the UP group, the amplitudes of the somatosensory threshold are also 0.3 to 0.7 mA, and it depends on each subject. AP increases by 4 mmHg from that in the supine position without GVS at the onset of HUT. Similarly, the increase in AP is augmented with applying subsensory GVS. Again, subsensory GVS with an amplitude of 0.1 mA below the somatosensory threshold enhanced AP control at the onset

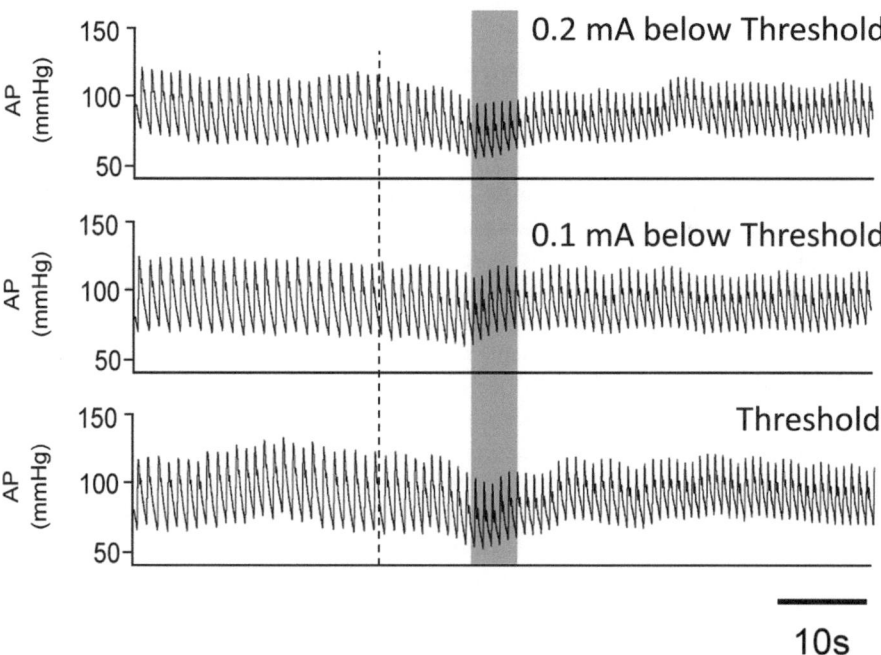

Fig. 3.4 Arterial pressure (AP) during 60° of head-up tilt (HUT) with galvanic vestibular stimulation (GVS). Applying GVS with amplitudes of 0.2 mA below the somatosensory threshold (first panel) and with amplitude of the threshold (third panel), AP is decreased at the onset of HUT (shaded part). However, applying GVS with amplitudes of 0.1 mA below the somatosensory threshold (second panel), AP is maintained

of HUT. As the phenomenon of stochastic resonance is used for the rehabilitation of sprains [41] and treatment of neuropathies [43], subsensory GVS may be used to prevent orthostatic hypotension derived from vestibular dysfunction.

3.4 Orthostatic Hypotension

As noted in this chapter, AP is controlled, most importantly, via the vestibular system at the onset of HUT or standing, and via baroreflex control after the onset and during standing. Thus, considering the pathophysiology of orthostatic hypotension, the mechanism should be divided if it is derived via vestibular dysfunction or decompensation of baroreflex control. In general, analysis and evaluation of orthostatic hypotension, as typified by the Schellong test, are performed more than 10 min after the onset of standing [44–47]. The orthostatic challenge using LBNP is useful for evaluating baroreflex according to blood shift and simulates orthostatic hypotension during prolonged standing [48, 49]. However, it is hard to simulate and to evaluate orthostatic hypotension at the onset of standing since LBNP usually do not change the vestibular inputs. In healthy individuals, a large initial decrease of AP upon standing was observed with initial orthostatic dizziness, one of the vestibular-related symptoms [50, 51]. Fifty percent of patients with vestibular-related impairment, aged ≥65 years, experienced orthostatic hypotension immediately on standing [52].

Orthostatic hypotension is often observed after exposure to microgravity conditions [53]. The vestibular and related functions are highly plastic or adaptive to the changing environment. Thus, the orthostatic hypotension after spaceflights might be related to the changes in vestibular function. After 4–6 months of stay in the international space station, we have presented that astronauts significantly decreased AP at the onset of HUT compared to that before launch [54]. This decrease is considered due to the vestibular function adapted to the microgravity environment. The vestibular function might have changed so that it does not react to gravity change [55]. AP was recovered after the decrease; thus, baroreflex might be maintained even after long-term exposure to microgravity. During HUT, after the long exposure to microgravity, subsensory GVS did not significantly raise AP at the onset of HUT. Thus, the subsensory GVS might be effective for an increase in vestibular sensitivity but not for the changed reflex pathway. Otherwise, the effects of changed cardiovascular function such as cardiac atrophy during exposure to microgravity might have a greater effect on AP control [56].

In children and adolescents, instantaneous orthostatic hypotension (INOH) has been reported as a new entity of orthostatic intolerance in subjects who underwent rapid standing as an orthostatic stress test. Systolic blood pressure in the patients decreased by 55 mmHg at the onset of standing, as seen in the response with GVS [57, 58]. Thus, clinical orthostatic hypotension at the onset of standing is considered to be related to vestibular-related dysfunction. The pathophysiology of INOH is still unclear, but misalignment of vestibule-cardiovascular reflex in growing children might be involved in the hypotension, as shown in this chapter.

In any case, the effects of a longer period of GVS for the vestibular-related AP control in aged or young subjects and astronauts are interesting since the stimulation may alter the vestibular and the related function due to plasticity.

3.5 Conclusion

The vestibular system senses a change in the direction of gravity for the body, probably increases sympathetic nerve activity, and reflexively induces vasoconstriction in advance to avoid a decrease in AP due to footward blood shift. However, AP is still decreased or increased with over-excitation since this control is a kind of feedforward control and is not induced by monitoring AP. After that, baroreflex, the feedback control, precisely controls AP to the appropriate value by monitoring AP. Orthostatic hypotension at the onset of standing is considered to be involved in vestibular-related autonomic control of AP.

References

1. Watenpaugh DE, Hargens AR. The cardiovascular system in microgravity. In: Fregly MJ, Blatteis CM, editors. Handbook of physiology, the graviational environment. Rockville, MD: American Physiological Society; 1996. p. 631–74.
2. Dampney RA, Stella A, Golin R, Zanchetti A. Vagal and sinoaortic reflexes in postural control of circulation and renin release. Am J Physiol. 1979;237(2):H146–52.
3. Sato T, Kawada T, Sugimachi M, Sunagawa K. Bionic technology revitalizes native baroreflex function in rats with baroreflex failure. Circulation. 2002;106(6):730–4.
4. Gauer OH, Thron HC. Postural changes in circulation. In: Hamilton WF, Dow P, editors. Handbook of physiology, circulation. Rockville, MD: American Physiological Society; 1965. p. 2409–39.
5. Mayerson HS. Influence of posture on blood flow in the dog. Am J Physiol. 1942;136:381–5.
6. Roddie IC, Shepherd JT, Whelan RF. Reflex changes in vasoconstrictor tone in human skeletal muscle in response to stimulation of receptors in a low-pressure area of the intrathoracic vascular bed. J Physiol. 1957;139(3):369–76.
7. Shepherd JT. Role of the veins in the circulation. Circulation. 1966;33(3):484–91.
8. Sjostrand T. Circulatory control via vagal afferents. V. Impairment of the circulatory adjustment to hemorrhage by vagal deafferentation and prolonged hypotension. Acta Physiol Scand. 1973;87(2):228–39.
9. Kamiya A, Michikami D, Fu Q, Iwase S, Hayano J, Kawada T, et al. Pathophysiology of orthostatic hypotension after bed rest: paradoxical sympathetic withdrawal. Am J Physiol Heart Circ Physiol. 2003;285(3):H1158–67.
10. Kingma H, Gauchard GC, de Waele C, van Nechel C, Bisdorff A, Yelnik A, et al. Stocktaking on the development of posturography for clinical use. J Vestib Res. 2011;21(3):117–25.
11. Massion J. Postural control system. Curr Opin Neurobiol. 1994;4(6):877–87.
12. Perrin P, Mallinson A, Van Nechel C, Peultier-Celli L, Petersen H, Magnusson M, et al. Defining clinical-posturographic and intra-posturographic discordances: what do these two concepts mean? J Int Adv Otol. 2018;14(1):127–9.
13. Speers RA, Kuo AD, Horak FB. Contributions of altered sensation and feedback responses to changes in coordination of postural control due to aging. Gait Posture. 2002;16(1):20–30.

14. Gotoh TM, Fujiki N, Matsuda T, Gao S, Morita H. Roles of baroreflex and vestibulosympathetic reflex in controlling arterial blood pressure during gravitational stress in conscious rats. Am J Physiol Regul Integr Comp Physiol. 2004;286(1):R25–30.
15. Tanaka K, Gotoh TM, Awazu C, Morita H. Roles of the vestibular system in controlling arterial pressure in conscious rats during a short period of microgravity. Neurosci Lett. 2006;397(1–2):40–3.
16. Abe C, Tanaka K, Awazu C, Morita H. Impairment of vestibular-mediated cardiovascular response and motor coordination in rats born and reared under hypergravity. Am J Physiol Regul Integr Comp Physiol. 2008;295(1):R173–80.
17. Hall JE. Cortical and brain stem control of motor function. In: Hall JE, editor. Textbook of medical physiology. Philadelphia: Elsevier; 2015. p. 707–19.
18. Ray CA, Hume KM, Shortt TL. Skin sympathetic outflow during head-down neck flexion in humans. Am J Physiol. 1997;273(3 Pt 2):R1142–6.
19. Ray CA, Hume KM. Neck afferents and muscle sympathetic activity in humans: implications for the vestibulosympathetic reflex. J Appl Physiol (1985). 1998;84(2):450–3.
20. Ray CA. Interaction of the vestibular system and baroreflexes on sympathetic nerve activity in humans. Am J Physiol Heart Circ Physiol. 2000;279(5):H2399–404.
21. Iwase S, Mano T, Cui J, Kitazawa H, Kamiya A, Miyazaki S, et al. Sympathetic outflow to muscle in humans during short periods of microgravity produced by parabolic flight. Am J Physiol. 1999;277(2 Pt 2):R419–26.
22. Iwata C, Abe C, Tanaka K, Morita H. Role of the vestibular system in the arterial pressure response to parabolic-flight-induced gravitational changes in human subjects. Neurosci Lett. 2011;495(2):121–5.
23. Abe C, Tanaka K, Iwata C, Morita H. Vestibular-mediated increase in central serotonin plays an important role in hypergravity-induced hypophagia in rats. J Appl Physiol. 2010;109(6):1635–43.
24. Bolton PS, Wardman DL, Macefield VG. Absence of short-term vestibular modulation of muscle sympathetic outflow, assessed by brief galvanic vestibular stimulation in awake human subjects. Exp Brain Res. 2004;154(1):39–43.
25. Goldberg JM, Smith CE, Fernandez C. Relation between discharge regularity and responses to externally applied galvanic currents in vestibular nerve afferents of the squirrel monkey. J Neurophysiol. 1984;51(6):1236–56.
26. Kim J, Curthoys IS. Responses of primary vestibular neurons to galvanic vestibular stimulation (GVS) in the anaesthetised Guinea pig. Brain Res Bull. 2004;64(3):265–71.
27. Watson SR, Brizuela AE, Curthoys IS, Colebatch JG, MacDougall HG, Halmagyi GM. Maintained ocular torsion produced by bilateral and unilateral galvanic (DC) vestibular stimulation in humans. Exp Brain Res. 1998;122(4):453–8.
28. Wardman DL, Fitzpatrick RC. What does galvanic vestibular stimulation stimulate? Adv Exp Med Biol. 2002;508:119–28.
29. Larsen PN, Moesgaard F, Madsen P, Pedersen M, Secher NH. Subcutaneous oxygen and carbon dioxide tensions during head-up tilt-induced central hypovolaemia in humans. Scand J Clin Lab Invest. 1996;56(1):17–24.
30. Abe C, Tanaka K, Awazu C, Morita H. Strong galvanic vestibular stimulation obscures arterial pressure response to gravitational change in conscious rats. J Appl Physiol. 2008;104(1):34–40.
31. Fitzpatrick RC, Day BL. Probing the human vestibular system with galvanic stimulation. J Appl Physiol. 2004;96(6):2301–16.
32. Carter JR, Ray CA. Sympathetic responses to vestibular activation in humans. Am J Physiol Regul Integr Comp Physiol. 2008;294(3):R681–8.
33. Kaufmann H, Biaggioni I, Voustianiouk A, Diedrich A, Costa F, Clarke R, et al. Vestibular control of sympathetic activity. An otolith-sympathetic reflex in humans. Exp Brain Res. 2002;143(4):463–9.
34. Tanaka K, Abe C, Sakaida Y, Aoki M, Iwata C, Morita H. Subsensory galvanic vestibular stimulation augments arterial pressure control upon head-up tilt in human subjects. Auton Neurosci. 2012;166(1–2):66–71.

35. Tanaka K, Abe C, Awazu C, Morita H. Vestibular system plays a significant role in arterial pressure control during head-up tilt in young subjects. Auton Neurosci. 2009;148(1–2):90–6.
36. Ray CA, Monahan KD. Aging attenuates the vestibulosympathetic reflex in humans. Circulation. 2002;105(8):956–61.
37. Engstrom H, Ades HW, Engstrom B, Gilchrist D, Bourne G. Structural changes in the vestibular epithelia in elderly monkeys and humans. Adv Otorhinolaryngol. 1977;22:93–110.
38. Rosenhall U, Rubin W. Degenerative changes in the human vestibular sensory epithelia. Acta Otolaryngol. 1975;79(1–2):67–80.
39. Zeng FG, Fu QJ, Morse R. Human hearing enhanced by noise. Brain Res. 2000;869(1–2):251–5.
40. Collins AT, Blackburn JT, Olcott CW, Miles J, Jordan J, Dirschl DR, et al. Stochastic resonance electrical stimulation to improve proprioception in knee osteoarthritis. Knee. 2010;18:317.
41. Ross SE, Arnold BL, Blackburn JT, Brown CN, Guskiewicz KM. Enhanced balance associated with coordination training with stochastic resonance stimulation in subjects with functional ankle instability: an experimental trial. J Neuroeng Rehabil. 2007;4:47.
42. Tanaka K, Ito Y, Ikeda M, Katafuchi T. RR interval variability during galvanic vestibular stimulation correlates with arterial pressure upon head-up tilt. Auton Neurosci. 2014;185:100–6.
43. Dhruv NT, Niemi JB, Harry JD, Lipsitz LA, Collins JJ. Enhancing tactile sensation in older adults with electrical noise stimulation. Neuroreport. 2002;13(5):597–600.
44. Cooper VL, Hainsworth R. Effects of head-up tilting on baroreceptor control in subjects with different tolerances to orthostatic stress. Clin Sci. 2002;103(3):221–6.
45. Rieckert H. Orthostatic hypotension: how to avoid it during antihypertensive therapy. Am J Hypertens. 1996;9(11):155S–9S.
46. Winker R, Fruhwirth M, Saul P, Rudiger HW, Pezawas T, Schmidinger H, et al. Prolonged asystole provoked by head-up tilt testing. Clin Res Cardiol. 2006;95(1):42–7.
47. Rutan GH, Hermanson B, Bild DE, Kittner SJ, LaBaw F, Tell GS. Orthostatic hypotension in older adults. The cardiovascular health study. CHS Collaborative Research Group. Hypertension. 1992;19(6 Pt 1):508–19.
48. Convertino VA. Aerobic fitness, endurance training, and orthostatic intolerance. Exerc Sport Sci Rev. 1987;15:223–59.
49. Convertino VA, Sather TM. Vasoactive neuroendocrine responses associated with tolerance to lower body negative pressure in humans. Clin Physiol. 2000;20(3):177–84.
50. Dambrink JH, Imholz BP, Karemaker JM, Wieling W. Postural dizziness and transient hypotension in two healthy teenagers. Clin Auton Res. 1991;1(4):281–7.
51. Wieling W, Shepherd JT. Initial and delayed circulatory responses to orthostatic stress in normal humans and in subjects with orthostatic intolerance. Int Angiol. 1992;11(1):69–82.
52. Ahearn DJ, Umapathy D. Vestibular impairment in older people frequently contributes to dizziness as part of a geriatric syndrome. Clin Med. 2015;15(1):25–30.
53. Buckey JC Jr, Lane LD, Levine BD, Watenpaugh DE, Wright SJ, Moore WE, et al. Orthostatic intolerance after spaceflight. J Appl Physiol (1985). 1996;81(1):7–18.
54. Morita H, Abe C, Tanaka K. Long-term exposure to microgravity impairs vestibulo-cardiovascular reflex. Sci Rep. 2016;6:33405.
55. Hallgren E, Migeotte PF, Kornilova L, Deliere Q, Fransen E, Glukhikh D, et al. Dysfunctional vestibular system causes a blood pressure drop in astronauts returning from space. Sci Rep. 2015;5:17627.
56. Levine BD, Pawelczyk JA, Ertl AC, Cox JF, Zuckerman JH, Diedrich A, et al. Human muscle sympathetic neural and haemodynamic responses to tilt following spaceflight. J Physiol. 2002;538(Pt 1):331–40.
57. Stewart JM, Weldon A. Inappropriate early hypotension in adolescents: a form of chronic orthostatic intolerance with defective dependent vasoconstriction. Pediatr Res. 2001;50(1):97–103.
58. Tanaka H, Yamaguchi H, Matushima R, Tamai H. Instantaneous orthostatic hypotension in children and adolescents: a new entity of orthostatic intolerance. Pediatr Res. 1999;46(6):691–6.

Chapter 4
Aging and Biological Rhythms

Dominika Kanikowska

Abstract This chapter is intended to clarify the significant roles of biological rhythms in aging.

People worldwide are living longer. As people age, they are more likely to experience several diseases related to advanced years. Older adults often suffer from circadian rhythm disturbances. Many aging-related diseases, such as Alzheimer's disease, Parkinson's disease, and malignancy, have a complex connection with biological rhythm. First, this chapter introduces the essential background regarding chronobiology. Second, it summarized the roles of biological rhythms in aging and aging-related diseases.

Keywords Biological rhythms · Aging · Chronotype · Clock gene · Melatonin · Cortisol · Sleep · Alzheimer's disease · Parkinson's disease

4.1 General Principles of Chronobiology

Physiological and behavioural processes of mammalian organisms follow rhythmic oscillations and show daily and seasonal rhythms, which reflect individual behaviours, such as, e.g. activity, sleep (called exogenous components of the daily rhythm), and internal "body clock" (called endogenous component). All organisms have evolved an internal timing system consisting of self-sustained oscillators reset by various synchronizers. Synchronization of the body clock to a 24-h day is achieved by "zeitgebers" (German: "time-giver"; a word Jurgen Aschoff

D. Kanikowska (✉)
Department of Pathophysiology, Poznan University of Medical Sciences, Poznan, Poland
e-mail: dkanikowska@ump.edu.pl

(1913–1989) contributed to the English language), rhythmic environmental influences [1]. The natural light-dark cycle is an essential factor (acting as a zeitgeber) influencing body temperature and hormonal rhythms [2]. The neuronal pathway transmitting light to the brain is known as the retinohypothalamic tract (RHT) [3].

All cells, tissues, and organs contain the genetic potential to manifest a clock, which gives rise to the concept of "peripheral" and "central" oscillators [1]. The central pacemaker in mammals is two hypothalamic clusters of neurons, the suprachiasmatic nuclei (SCN). SCN controls behavioural, metabolic, and physiological rhythms and synchronizes the peripheral oscillators [4, 5]. Whereas the rhythmicity of the SCN is maintained mainly through the light-dark cycle, that in peripheral oscillators is controlled through the rest-activity or feeding-fasting cycles (acting as zeitgebers) or rhythms in circulating hormones such as melatonin or cortisol [6]. The SCN gives rise to the endogenous component of an observed daily rhythm. The exogenous component of the daily rhythm corresponds to the environment and the individual's lifestyle. Typically, endogenous and exogenous components of a rhythm are in phase. However, after a time-zone transition or night work, they become desynchronized because the exogenous component adjusts faster with the individual's environment, and the SCN is delayed adjusting [1]. The circadian rhythm could be assessed by a single cosine analysis, where rhythm is described by the midline estimating statistic of rhythm (MESOR), amplitude, and acrophase. The MESOR is the mean of all values across the circadian rhythm. The amplitude is half the difference between the highest and the lowest points of the cosine function, best fitting the data. The acrophase represents the time point when the circadian cycle reaches the peak value [7].

4.2 Molecular Chronobiology

Greatly simplified, cellular clocks consist of transcriptional/translational autoregulatory feedback loops [8]. The main loop comprises the transcriptional activators CLOCK and BMAL 1 and their target genes (namely clock genes) Per (Period) and Cry (Cryptochrome). The transcription factors CLOCK:BMAL1 coordinate the rhythmic expression of the transcriptome and control the daily regulation of biological functions. These gene products accumulate and inhibit the transcription of CLOCK-BMAL 1. The feedback loop controlling BMAL 1 transcriptions is stable with inhibition by the nuclear receptors REV-ERBα and activation by ROR α and β. Importantly, REV-ERBα links metabolism to the clock system [9].

4.3 Periodic Fluctuation of Endocrine Secretion

The body clock is synchronized to a 24-h cycle, with the natural light-dark cycle playing a critical role. In the absence of time cues, the dominant component of this system is "free-runs" with a period of 24–25 h ("circadian," from the Latin: "about a day") [10].

Under normal conditions, the light-dark cycle acts harmoniously to entrain the body clock to a solar (24-h) day. In humans, this results in daily activity and night sleep, with the peak of body temperature being near the middle of daytime activity and its trough near the middle of night sleep. Also, the changes in light intensities during the daytime have influenced body temperature and melatonin level rhythm [11].

Depending on the length of the cycle of periodic fluctuation, ultradian rhythms (with cycles shorter than 20 h) and infradian rhythms (28 h or more) can also be distinguished.

Several hormones exhibit pulsatory/ ultradian oscillations (i.e. growth hormone) and infradian/seasonal fluctuations (i.e. cortisol, melatonin) [12].

Melatonin and cortisol are involved in the body's sleep-wake cycles and are essential components of the circadian rhythm. Melatonin is a hormone produced by the pineal gland and is related chemically to serotonin, a hormone whose secretion is connected with sleep and mood. Melatonin is the "hormone of darkness" because its secretion increases soon after the onset of darkness, and peaks of rhythm appear in the middle of the night and gradually fall during the second half of the night. Moreover, its secretion is suppressed by bright light. Furthermore, the SCN regulates the circadian production and secretion of melatonin [13]. Melatonin not only has an essential role in regulating circadian rhythms but also is an antioxidant and neuroprotector that may be important in aging and Alzheimer's disease (AD) [14].

Cortisol is a subsistence hormone that is fundamental to the maintenance of homeostasis. It's produced by the adrenal gland, mainly regulated by the hypothalamic-pituitary-adrenal (HPA) axis, and the secretion of cortisol provides a link between the body and mind. Cortisol secretion exhibits well-recognized circadian rhythmicity, and peaks of the rhythm appear in the early morning and gradually fall during the day and the first half of the night [13]. The cortisol concentration concentrations demonstrated differences between the seasons in terms of both average levels and daily rhythmicity [15].

The secretion of growth hormone (GH) and sex hormones show the most marked changes with aging. GH regulates many aspects of metabolism, and secretion is mainly associated with the deepest stage of sleep—slow-wave sleep (SWS) [16].

4.4 Chronotype

The circadian system determines the human chronotype. Humans vary in the time of day when they prefer to sleep and their length of sleep. The chronotype is a pattern of the circadian rhythm that determines the most optimal activity and rest time and represents the relationship between the phase of one's endogenous rhythm and external synchronizers. Chronotype is divided into morning ("lark"), evening ("owl"), and intermediate type [17].

4.5 Aging

Healthy senescence has been defined as developing and maintaining the functional ability that enables well-being in older age. According to the WHO, already, there are more than 1 billion people aged 60 years or older [18]. Additionally, life expectancy and the percentage of people older than 60 have continuously increased. Aging is associated with certain diseases, e.g., Alzheimer's (AD) and Parkinson's (PD), cardiovascular and malignancies [19]. It is not surprising, therefore, that biological rhythms also change with age and could result in the consequences of health problems [20].

4.5.1 Synchronization and Aging

Age-related disturbances have been noted in both zeitgebers and synchronization. SCN shows degenerative disturbances with aging, which could be related to a decrease in the number of vasopressin-expressing neurons observed in older people. In addition, the light-dark cycle, which represents the principal zeitgeber reaching SCN and pineal gland via the RHT path, could also be reduced in aged people [19]. One of the most common age-related diseases is cataracts, where the ability of the lens to transmit light progressively decreases, which could contribute to alterations in circadian synchronization [20].

4.5.2 Clock Genes and Aging

Biological rhythms are regulated at the transcriptional level by several interacting clock genes. Alterations in these clock genes' transcription and translation loops could alter periodic fluctuations of circadian and other rhythms.

The clock genes' circadian rhythm during aging is characterized by a decreased amplitude and alterations of acrophase [21]. In laboratory studies on animal models, different life span durations were observed for wide-type animals and clock genes knockout ones.

For example, the average lifespan of BMAL1-deficient mice is shorter than their wild-type controls. Moreover, BMAL1-deficient mice develop several age-related diseases, including sarcopenia, osteoporosis, and cataracts. Similar to BMAL1-deficient mice, CLOCK-knockout mice also have a lower average lifespan than wild-type mice and also develop more often age-specific pathology, such as cataracts. In the SCN and other brain areas of aged animals, Clock mRNA expression is significantly reduced. Per1, Per2 knockout mice develop features of premature aging, such as a decrease in fertility and loss of soft tissues [21].

4.5.3 Sleep-Wake Activity in Aging

Sleep is a complex neurological state, defined as unconsciousness from which some stimuli can arouse the person. Human sleep is divided into non-rapid eye movement (NREM) and rapid eye movement (REM) sleep. NREM sleep could be classified into four stages (namely stages I, II, III, and IV), representing successively deeper stages of sleep. During REM sleep, the brain is most active, where dreams usually occur. Alterations in the quality, quantity, and pattern of sleep result in sleep disorders. In most organisms (including humans), the cycle of sleeping and wakefulness is expressed as a 24-h entrained rhythm determined by the body clock. In particular, sleep propensity depends on the body clock and homeostatic sleep drive. SCN and the pineal gland ensure that sleep and wakefulness follow a circadian periodicity of nearly 24 h [22]. The timing of sleep propensity is associated with increased melatonin secretion and decreases in alertness and body temperature. Melatonin production decreases with age and in Alzheimer's disease. As well as with age, the decline in the amplitude of the circadian rhythm of the body temperature is observed. This decrease in melatonin production and the amplitude of 24 h body temperature rhythm is connected with insomnia in older patients. Moreover, age increases the difficulty of obtaining the recovery sleep needed following sleep deprivation, and older patients show difficulty sleeping at an adverse circadian phase [23].

4.5.4 Melatonin, Cortisol, and Aging

Melatonin and cortisol are crucial components of the circadian system. Both hormones' secretion is reported to decrease with the brain's aging, which may also be implicated in brain aging in the elderly.

The reason for the decline of melatonin secretion is possible by several factors, including calcification of the pineal gland, reduction in sensitivity and density of pineal receptors, and lower secretion of serotonin and norepinephrine. These hormones are essential for melatonin synthesis. Moreover, the acrophase of the melatonin rhythm occurred later in the elderly, and the acrophase of the circadian melatonin rhythm had a significant negative correlation with advancing age. Along with the acrophase, the amplitudes of melatonin showed a negative correlation with age and were significantly lower. In conclusion, in the elderly are observed lower activity of the pineal gland and delayed rise of the nocturnal peak in the daily melatonin rhythm [24].

Similar to melatonin, cortisol secretion also decreases with age. A decrease in cortisol secretion in aging people is related to several factors, including lower sensitivity and impairment of the HPA axis regulation, decreasing the secretion of pituitary adrenocorticotrophic hormone (ACTH) and the corticotropin-releasing hormone (CRH). Furthermore, when observing the daily cortisol rhythm in older people, the acrophase of the cortisol rhythm occurred earlier. Some studies suggested that the phase advance of the cortisol rhythm in the elderly may be linked to their "morningness." Along with the acrophase, the amplitudes of cortisol showed a negative correlation with age and were significantly lower. Furthermore, a decline in other hormones is observed during aging. The secretion of growth hormone (GH) and sex hormones show the most marked changes with aging. GH regulates many aspects of metabolism, and secretion is mainly associated with the deepest stage of sleep—slow-wave sleep (SWS). The lowered concentration of GH observed during sleep in the elderly is associated, among others, with shorter SWS [25].

Viewed together, this suggests that not only does the production of both melatonin and cortisol significantly decline with age, but also circadian rhythms change. Their respective acrophases move in opposite directions: earlier in the case of cortisol and later to melatonin [24]. As a final point, it can be concluded that aging subjects demonstrate some elements of desynchrony of the body clock, which can be explained by either the changes in the period of the circadian rhythms or the relationship between the zeitgeber and the body clock [26].

4.5.5 Body Temperature and Aging

In mammals, thermoregulatory mechanisms maintain a constant body temperature regardless of ambient temperature. In humans, physiologically, in the resting condition, body temperature shows a circadian rhythm with the lowest temperature between 01:00–06:00 h and the highest between 13:00–18:00 h [27]. The resting circadian rhythm of core temperature is caused mainly by heat loss and is mediated by cutaneous vasodilatation and sweating from the distal limbs. The maintenance of temperature regulations decreases as people age, and as a result, older people are more susceptible to cold. It is partly because they have more difficulty producing heat through muscle activity and shivering. In addition, cutaneous blood regulation and sweating are less active. Finally, older people are less active.

Moreover, the daily rhythm of body temperature has changed with age. The observed daily rhythm decreased in amplitude, increased in the day-to-day timing of the rhythm, and the acrophase of rhythm advanced [28].

4.5.6 Chronotype and Aging

Humans vary in the "morningness-eveningness" [17]. It depends not only on environmental factors but also on age. Moreover, chronotype changes with age; it changes earlier during the elderly and the latest during adolescence. Aging people (similar to children) are, on average, earlier chronotypes. People reach a maximum in "eveningness" at about 20, then become "morningness" again with increasing age.

Besides, it was shown that chronotypes differ across sex. Women tend to mature earlier than males in the changes of chronotype. Females reach their maximum in "eveningness" at about 19.5 years, whereas men at about 21 years old. However, this sex difference disappears, and females and men over 60 years of age are morning chronotypes on average [29].

4.5.7 Output

Several pathological conditions in older age include neurodegenerative diseases, most prominently Parkinson's and Alzheimer's diseases correlated with sleep disorders and circadian dysfunction [30]. In AD is observed circadian disorders, such as disturbances of melatonin and sleep-wake rhythms [14].

Melatonin production and daily rhythm are reduced or disappear, possibly due to dysfunction of pineal gland sympathetic regulation and RHT degeneration. AD patients showed an abnormal pattern of daily melatonin rhythm with low nocturnal melatonin levels and high daytime melatonin.

Sleep-wake rhythm disturbances, insomnia, and fragmented sleep-wake patterns also often occur in elderly AD patients, which could be consistent with impairment of melatonin rhythm and secretion. Furthermore, they often presented circadian system-related behavioural symptoms, such as daytime agitation and nightly restlessness.

In Parkinson's disease, patients often suffer from sleep-wake disturbances, including changes in sleep timing and extreme daytime sleepiness [31]. Daily melatonin and cortisol rhythms showed reduced amplitude. Moreover, other alterations in daily rhythms are observed in PD patients, including nocturnal hypertension, which is connected with a reversed circadian rhythm in blood pressure, with a high night blood pressure, higher or equal to daytime levels. In PD, patients have also observed disruptions in circadian thermoregulation, with a reduction in the amplitude of the body temperature rhythm. In addition, in Parkinson's patients, compared to healthy individuals is observed altered and reduced night BMAL1 mRNA expression [32].

4.6 Conclusions

Biological rhythms play an essential role in coordinating various physiological and behavioural processes. Therefore, alterations in daily hormone patterns during aging are highly likely to contribute to the ethology of various pathologies in the elderly.

Viewed together, these suggest that not only is senescence characterized by disturbances of biological rhythms, but also age-related perturbation of circadian rhythms could impact the aging process. As a final point, biological rhythms may be implicated in the process of brain aging, and it emphasizes the importance of studying the role of biological rhythms in the aging process.

References

1. Reilly T, Peiser B. Seasonal variations in health-related human physical activity. Sports Med. 2006;36(6):473–85. https://doi.org/10.2165/00007256-200636060-00002.
2. Shanahan TL, Czeisler CA. Light exposure induces equivalent phase shifts of the endogenous circadian rhythms of circulating plasma melatonin and core body temperature in men. J Clin Endocrinol Metab. 1991;73(2):227–35. PMID: 1856258. https://doi.org/10.1210/jcem-73-2-227.
3. Berson DM, Dunn FA, Takao M. Phototransduction by retinal ganglion cells that set the circadian clock. Science. 2002;295(5557):1070–3. PMID: 11834835. https://doi.org/10.1126/science.1067262.
4. Welsh DK, Takahashi JS, Kay SA. Suprachiasmatic nucleus: cell autonomy and network properties. Annu Rev Physiol. 2010;72:551–77. PMID: 20148688; PMCID: PMC3758475. https://doi.org/10.1146/annurev-physiol-021909-135919.
5. Reppert SM, Weaver DR. Molecular analysis of mammalian circadian rhythms. Annu Rev Physiol. 2001;63:647–76. PMID: 11181971. https://doi.org/10.1146/annurev.physiol.63.1.647.
6. Balsalobre A, Brown SA, Marcacci L, Tronche F, Kellendonk C, Reichardt HM, Schütz G, Schibler U. Resetting of circadian time in peripheral tissues by glucocorticoid signaling. Science. 2000;289(5488):2344–7. PMID: 11009419. https://doi.org/10.1126/science.289.5488.2344.
7. Halberg F. Chronobiology. Annu Rev Physiol. 1969;31:675–725. PMID: 4885778. https://doi.org/10.1146/annurev.ph.31.030169.003331.
8. Albrecht U. Timing to perfection: the biology of central and peripheral circadian clocks. Neuron. 2012;74(2):246–60. PMID: 22542179. https://doi.org/10.1016/j.neuron.2012.04.006.
9. Bass J, Takahashi JS. Circadian integration of metabolism and energetics. Science. 2010;330(6009):1349–54. PMID: 21127246; PMCID: PMC3756146. https://doi.org/10.1126/science.1195027.
10. Halberg F. Temporal coordination of physiologic function. Cold Spring Harb Symp Quant Biol. 1960;25:289–310. PMID: 13710673. https://doi.org/10.1101/sqb.1960.025.01.031.
11. Kanikowska D, Hirata Y, Hyun K, Tokura H. Acute phase proteins, body temperature and urinary melatonin under the influence of bright and dim light intensities during the daytime. J Physiol Anthropol Appl Human Sci. 2001;20(6):333–8. PMID: 11840685. https://doi.org/10.2114/jpa.20.333.
12. Hastings M, O'Neill JS, Maywood ES. Circadian clocks: regulators of endocrine and metabolic rhythms. J Endocrinol. 2007;195(2):187–98. PMID: 17951531. https://doi.org/10.1677/JOE-07-0378.
13. Haus E. Chronobiology in the endocrine system. Adv Drug Deliv Rev. 2007;59(9–10):985–1014. Epub 2007 Jul 14. PMID: 17804113. https://doi.org/10.1016/j.addr.2007.01.001.
14. Wu YH, Swaab DF. The human pineal gland and melatonin in aging and Alzheimer's disease. J Pineal Res. 2005;38(3):145–52. PMID: 15725334. https://doi.org/10.1111/j.1600-079X.2004.00196.x.
15. Kanikowska D, Roszak M, Rutkowski R, Sato M, Sikorska D, Orzechowska Z, Bręborowicz A, Witowski J. Seasonal differences in rhythmicity of salivary cortisol in healthy adults. J Appl Physiol (1985). 2019;126(3):764–70. Epub 2019 Jan 31. PMID: 30702977. https://doi.org/10.1152/japplphysiol.00972.2018.
16. Gronfier C, Luthringer R, Follenius M, Schaltenbrand N, Macher JP, Muzet A, Brandenberger G. A quantitative evaluation of the relationships between growth hormone secretion and delta wave electroencephalographic activity during normal sleep and after enrichment in delta waves. Sleep. 1996;19(10):817–24. PMID: 9085491. https://doi.org/10.1093/sleep/19.10.817.
17. Horne JA, Ostberg O. A self-assessment questionnaire to determine morningness-eveningness in human circadian rhythms. Int J Chronobiol. 1976;4(2):97–110.
18. https://www.who.int/news-room/fact-sheets/detail/ageing-and-health.

19. Le Couteur DG, Thillainadesan J. What is an aging-related disease? An epidemiological perspective. J Gerontol A Biol Sci Med Sci. 2022;77:2168. Epub ahead of print. PMID: 35167685. https://doi.org/10.1093/gerona/glac039.
20. Turek FW, Van Reeth O. Comprehensive physiology circadian rhythms. Supplement 14. In: Handbook of physiology, environmental physiology; 2011. https://doi.org/10.1002/cphy.cp040258.
21. Froy O. Circadian aspects of energy metabolism and aging. Ageing Res Rev. 2013;12(4):931–40. Epub 2013 Sep 25. PMID: 24075855. https://doi.org/10.1016/j.arr.2013.09.002.
22. Münch MY, Cain SW, Duffy JF. Biological rhythms workshop IC: sleep and rhythms. Cold Spring Harb Symp Quant Biol. 2007;72:35–46. PMID: 18419261. https://doi.org/10.1101/sqb.2007.72.065.
23. Czeisler CA, Gooley JJ. Sleep and circadian rhythms in humans. Cold Spring Harb Symp Quant Biol. 2007;72:579–97. PMID: 18419318. https://doi.org/10.1101/sqb.2007.72.064.
24. Sharma M, Palacios-Bois J, Schwartz G, Iskandar H, Thakur M, Quirion R, Nair NP. Circadian rhythms of melatonin and cortisol in aging. Biol Psychiatry. 1989;25(3):305–19. PMID: 2914154. https://doi.org/10.1016/0006-3223(89)90178-9.
25. Pace-Schott EF, Spencer RM. Age-related changes in the cognitive function of sleep. Prog Brain Res. 2011;191:75–89. PMID: 21741545. https://doi.org/10.1016/B978-0-444-53752-2.00012-6.
26. Terzibasi-Tozzini E, Martinez-Nicolas A, Lucas-Sánchez A. The clock is ticking. Ageing of the circadian system: from physiology to cell cycle. Semin Cell Dev Biol. 2017;70:164–76. Epub 2017 Jun 16. PMID: 28630025. https://doi.org/10.1016/j.semcdb.2017.06.011.
27. Minors DS, Waterhouse JM. The sleep-wakefulness rhythm, exogenous and endogenous factors (in man). Experientia. 1984;40(5):410–6. PMID: 6373354. https://doi.org/10.1007/BF01952373.
28. Gubin DG, Gubin GD, Waterhouse J, Weinert D. The circadian body temperature rhythm in the elderly: effect of single daily melatonin dosing. Chronobiol Int. 2006;23(3):639–58. PMID: 16753947. https://doi.org/10.1080/07420520600650612.
29. Roenneberg T, Merrow M. Entrainment of the human circadian clock. Cold Spring Harb Symp Quant Biol. 2007;72:293–9. PMID: 18419286. https://doi.org/10.1101/sqb.2007.72.043.
30. Hofman MA, Boer GJ, Holtmaat GD, EJW VS, Verhaagenand J, Swaab DF, editors. Progress in brain research, vol. 138. Amsterdam: Elsevier; 2002.
31. Hunt J, Coulson EJ, Rajnarayanan R, Oster H, Videnovic A, Rawashdeh O. Sleep and circadian rhythms in Parkinson's disease and preclinical models. Mol Neurodegener. 2022;17(1):2. PMID: 35000606; PMCID: PMC8744293. https://doi.org/10.1186/s13024-021-00504-w.
32. Ding H, Liu S, Yuan Y, Lin Q, Chan P, Cai Y. Decreased expression of Bmal2 in patients with Parkinson's disease. Neurosci Lett. 2011;499(3):186–8. Epub 2011 May 30. PMID: 21658431. https://doi.org/10.1016/j.neulet.2011.05.058.

Chapter 5
Development of 3D Contents for Extracting Aging Characteristics Using Electrical Physiology

Fumiya Kinoshita, Kikuo Ito, and Hiroki Takada

Abstract Sports vision, which has been attracting attention recently, has been incorporated into training practices that aim to maintain and recover visual functions. Not only is it effective in saving space, but it is also effective in maintaining cognitive function, which has been stated in numerous neuroscientific discussions. In this chapter, we use electroencephalograms that were measured during certain tasks using the visuospatial brain gymnastic 3D program that we are developing. Subsequently, we present an overview of the differences between the feature values of each age group.

Keywords Mild cognitive impairment (MCI) · Visuospatial cognitive impairment · Sports vision · 3D · Electroencephalogram

5.1 Aging Society and Dementia

The percentage of people aged 65 and older in the total national population is defined as the aging rate, which is commonly used as an indicator of the "aging of a country." As of September 2022, Japan's aging rate was 29.1%, which was the

---

F. Kinoshita (✉)
Graduate School of Engineering, Toyama Prefectural University,
Kurokawa, Imizu-shi, Toyama, Japan
e-mail: f.kinoshita@pu-toyama.ac.jp

K. Ito
Neurosky Japan Inc, Chuo-ku, Tokyo, Japan

H. Takada
Department of Human and Artificial Intelligence Systems, Graduate School of Engineering, University of Fukui, Fukui, Fukui, Japan

© The Author(s), under exclusive license to Springer Nature Singapore Pte Ltd. 2024, corrected publication 2024
T. Shiozawa et al. (eds.), *Gerontology as an Interdisciplinary Science*, Current Topics in Environmental Health and Preventive Medicine, https://doi.org/10.1007/978-981-97-2712-4_5

highest in the world, indicating the seriousness of aging in Japan [1]. The main problem in an aging society is the increase in social security and long-term care insurance costs associated with the increase in elderly people. Efforts to maintain and improve the health prospects of the elderly are important to prevent the increase in medical care costs. Dementia is one of the three major diseases that increase the need for nursing care for the elderly. Dementia is defined as a "condition in which once-normal cognitive function persistently decreases owing to acquired brain disorders and interferes with daily and social life." However, a diagnostic method has not yet been strictly determined, and it is impossible to completely cure dementia even with current medicine [2]. Dementia is thought to induce increased physical, psychological, and economic burdens not only on the affected individual but also on the family and community. Therefore, there are increasing expectations for methods to prevent dementia and delay decreased cognitive function.

Early detection of mild cognitive impairment (MCI), which is at the boundary between a healthy cognitive state and dementia, is important for delaying decreased cognitive function. MCI is a concept established by Ron Petersen in the United States in 1999, which is described as a state that "cannot be described as normal but also cannot be described as dementia" [3]. Subsequently, patients diagnosed with MCI are at high risk of developing dementia; thus, MCI is considered to be a prior stage to dementia. Originally, MCI was a concept that emphasized memory impairment. However, it has recently been classified into "amnestic MCI," which is mainly characterized by memory impairment, and "non-amnestic MCI," which is characterized by impairment in performance, attention, language, and visuospatial cognition. A non-impaired state is when an individual can "live daily life independently." Therefore, a diagnosis needs to involve collecting testimony from a person who knows the diagnosed individual well and can state that the "current cognitive level of the individual in question has deteriorated when compared to the level that the individual in question had maintained beforehand." Objective cognitive function tests are important in diagnosing MCI, and multifaceted evaluations of multiple cognitive domains, such as memory, language, executive function, visuospatial cognition, and attention, are needed. A widely used evaluation framework for cognitive function is the Montreal Cognitive Assessment (MoCA) because it is thought to be superior in detecting MCI compared to other methods [4]. However, although the screening test using memory load is effective in detecting amnestic MCI, it does not have high detection accuracy for non-amnestic MCI. The progression rate from MCI to dementia is approximately 10% per year, which is not a low value [5]. Thus, establishing a test method that can detect non-amnestic MCI at an early stage is important.

5.2 Early Detection and Countermeasures for Mild Dementia

Early symptoms of Alzheimer's dementia, such as difficulty in drawing figures, getting lost while driving, and being unable to park in a garage, are known to appear even without visual impairment [6]. These symptoms, collectively referred to as visuospatial cognitive impairment, are important clinical manifestations for early diagnosis. There are two pathways for visual information processing in the cerebral cortex [7]. One is the "ventral visual pathway," which processes information for identifying objects. The ventral visual pathway starts from the primary visual cortex and passes through the temporal cortex, arriving at the prefrontal cortex. The other is the "dorsal visual pathway," which processes information such as the position and depth of objects. The dorsal visual pathway also originates from the primary visual cortex; however, this visual pathway leads through the parietal cortex, arriving at the prefrontal cortex. The visual information received from these two pathways converge again in the prefrontal cortex, where visual information processing is conducted. In Alzheimer's dementia, not only is there damage around the hippocampus, but also blood flow in the parietal-occipital lobe is reduced from an early stage of the disease [8]. In other words, the decreased blood flow in the parietal-occipital lobe seen in Alzheimer's dementia affects the dorsal visual pathway dysfunction. Accordingly, it is thought that this becomes the lesion responsible for the illness, with clinical symptoms, such as "I can understand what the object is, but I can't understand spatial information such as its exact position and depth," appearing. Visuospatial cognitive impairment is a clinical manifestation that is not only of Alzheimer's dementia but also of MCI. Therefore, if visuospatial cognitive function can be evaluated in a quantitative manner, then it will be useful for early detection of MCI. Therefore, our research group is focused on the early detection of MCI by measuring brain processes related to stereoscopic vision [9–11]. In the following section, we will outline the difference in feature values for each age group using electroencephalograms (EEG) measured during certain tasks using the visuospatial brain exercise 3D program we are developing.

The visuospatial brain exercise 3D program we are developing consists of four 3D video contents. For content (1), a pendulum sphere that moves along the depth direction (i.e., back and forth) is tracked with central vision; subsequently, training is conducted 30 times to predict the time when it will stop temporarily at the closest point (Fig. 5.1a). The speed of the pendulum increases every 10 swings. Here, the prediction and assessment accuracy in the central vision is evaluated. For content (2), training is conducted 15 times to accurately recognize a moving sphere's color change (Fig. 5.1b). The sphere's speed increases every five moves. As the training progresses, the sphere's movement spreads across the entire screen. Thus, the situation must be assessed using peripheral vision. This is closely related to the ability to adapt to constantly changing traffic circumstances in daily life. For content (3), short-term memory training is conducted 15 times by correctly answering the types and numbers of displayed fruits of different sizes (Fig. 5.1c). The display time is shortened every five moves. Short-term memory is strengthened by converting displayed fruits into hiragana and entering the corresponding numbers. For content (4), the order of the objects placed in the depth direction is stated, and color and spatial perception training is conducted 15 times (Fig. 5.1d). The number and complexity of the objects arranged increases every 5 times. In other words, settings with the same color but different shapes and levels of brightness are added to the assessment conditions. Similar to content (2), spatial cognition is essential in daily life. Contents (3) and (4) are mediated by the perception of size and color constancy, which also serves as training for higher functions.

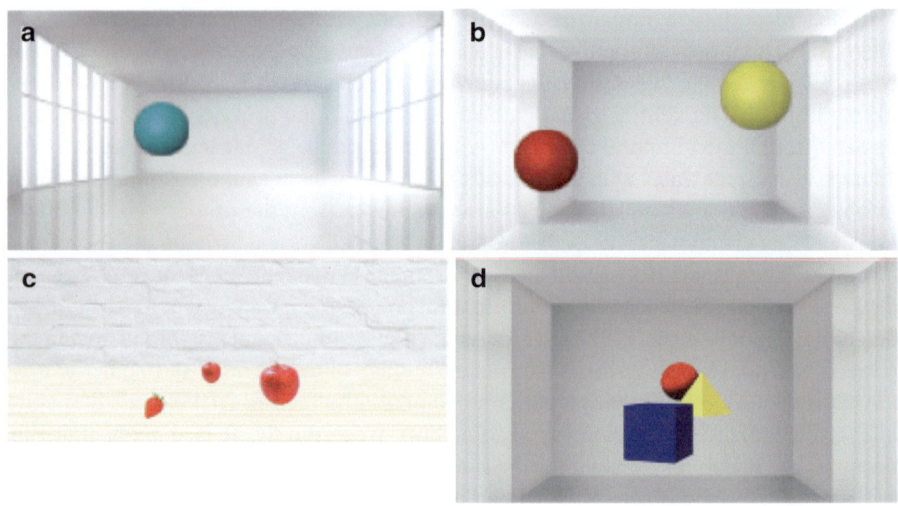

Fig. 5.1 Contents of the visuospatial brain exercise 3D program: (**a**) prediction training using a simple pendulum, (**b**) discrimination training for color changes in moving spheres, (**c**) short-term memory training for five kinds of fruits and their numbers, and (**d**) color constancy training

5.3 Validation

We asked 160 healthy subjects (mean ± standard deviation: 43.10 ± 18.92) to conduct the content (3) of the visuospatial brain exercise 3D program while we measured their EEGs. The simple electroencephalograph Mind Wave Mobile (Neurosky Co., Ltd. (Tokyo)) was used as the measurement device. The sampling frequency of this measurement device was 128 Hz, the EEG electrode was Fp1, and the reference electrode was the left earlobe. During the analysis, the window width of 1 s was divided every 300 ms, and frequency analysis was conducted using the fast Fourier transform (FFT) for each divided time series. The power spectrum of each acquired time-series data is classified into seven bands (θ, low-α, high-α, low-β, high-β, low-γ, and mid-γ); accordingly, corresponding integrated power spectral density (PSD) values were calculated. In this experiment, the θ wave band was 3.5–6.75 Hz, low-α wave band was 7.5–9.25 Hz, high-α wave band was 10–11.75 Hz, low-β wave band was 13–16.75 Hz, high-β wave band was 18–29.75 Hz, low-γ wave band was 31–39.75 Hz, and mid-γ wave band was 41–49.75 Hz. The time to execute content (3) varied for each subject; therefore, we used the median value of the time-series data as the representative value for the subjects. Additionally, we excluded data corresponding to when the integrated PSD exceeded the mean ± standard deviation during the analysis to reduce noise, such as eye blinks and body movements. In this experiment, 160 subjects were classified by age group and score (percentage of correct answers) as reported in Tables 5.1 and 5.2, respectively; subsequently, statistical processing was conducted. For statistical processing, multiple comparisons were made based on conducting the Steel-Dwass test after the Kruskal-Wallis test. The significance level in this experiment was set to 0.05.

Table 5.1 Classification by age group

Group name	Age group	Number of subjects [people]
Group-A	≥65 years	30
Group-B	50–64 years	31
Group-C	40–49 years	26
Group-D	≤39 years	73

Table 5.2 Classification by score group

Group name	Age group	Number of subjects [people]
Group-I	≤75 points	19
Group-II	76–85 points	25
Group-III	≥86 points	116

5.4 Results

The subjects were classified into four age groups, as reported in Table 5.1; accordingly, statistical processing was conducted. The results are shown in Fig. 5.2. The average score for each group was 76 points for Group A, 90 points for Group B, 92 points for Group C, and 94 points for Group D (Fig. 5.2a). After conducting the Kruskal-Wallis test, multiple comparisons were made based on the results of the Steel-Dwass test. Consequently, significant differences were observed in all comparisons ($p < 0.05$). Next, we classified the power spectrum of each acquired time-series dataset into seven bands, calculated each integral PSD value, and conducted statistical processing. As a result, the low-α, high-α, low-β, high-β, and low-γ values (Fig. 5.2c–g) significantly decreased in Group D compared to those in Group A ($p < 0.05$). Additionally, the high-α, low-β, high-β, and low-γ values (Fig. 5.2d–g) significantly decreased in Group C compared to those in Group A ($p < 0.05$).

Next, the subjects were classified into three score-tier groups, as reported in Table 5.2; accordingly, statistical processing was conducted. The results are shown in Fig. 5.3. The mean age for each group was 67 years for Group I, 53 years for Group II, and 37 years for Group III (Fig. 5.3a). After conducting the Kruskal-Wallis test, multiple comparisons were made based on the results of the Steel-Dwass test. Consequently, significant differences were observed in all comparisons ($p < 0.05$). Next, we classified the power spectrum of each acquired time-series dataset into seven bands, calculated each integral PSD value, and conducted statistical processing. Results showed that the low-α, high-α, low-β, high-β, and low-γ values (Fig. 5.3c–g) significantly decreased in Group II compared to those in Group I ($p < 0.05$).

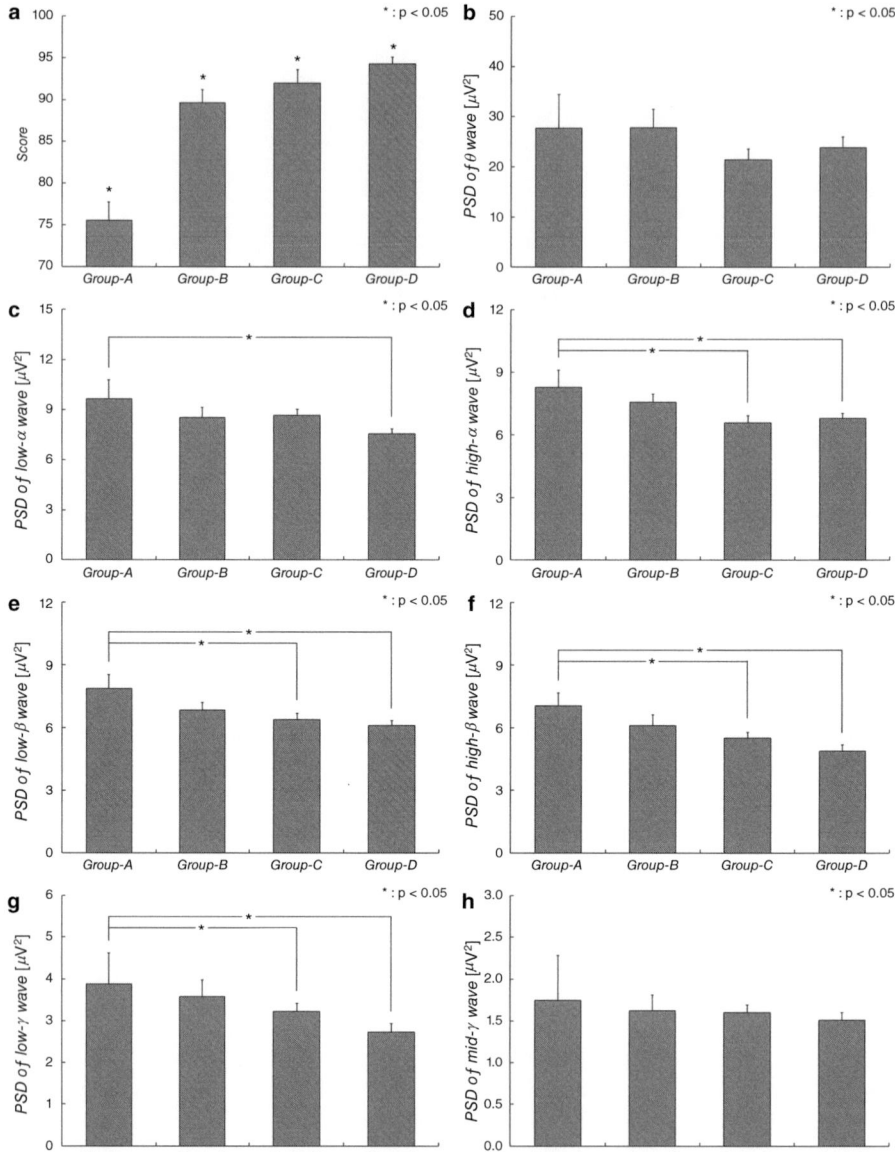

Fig. 5.2 Classification by age group (median ± SE): (**a**) score, (**b**) θ, (**c**) low-α, (**d**) high-α, (**e**) low-β, (**f**) high-β, (**g**) low-γ, and (**h**) mid-γ

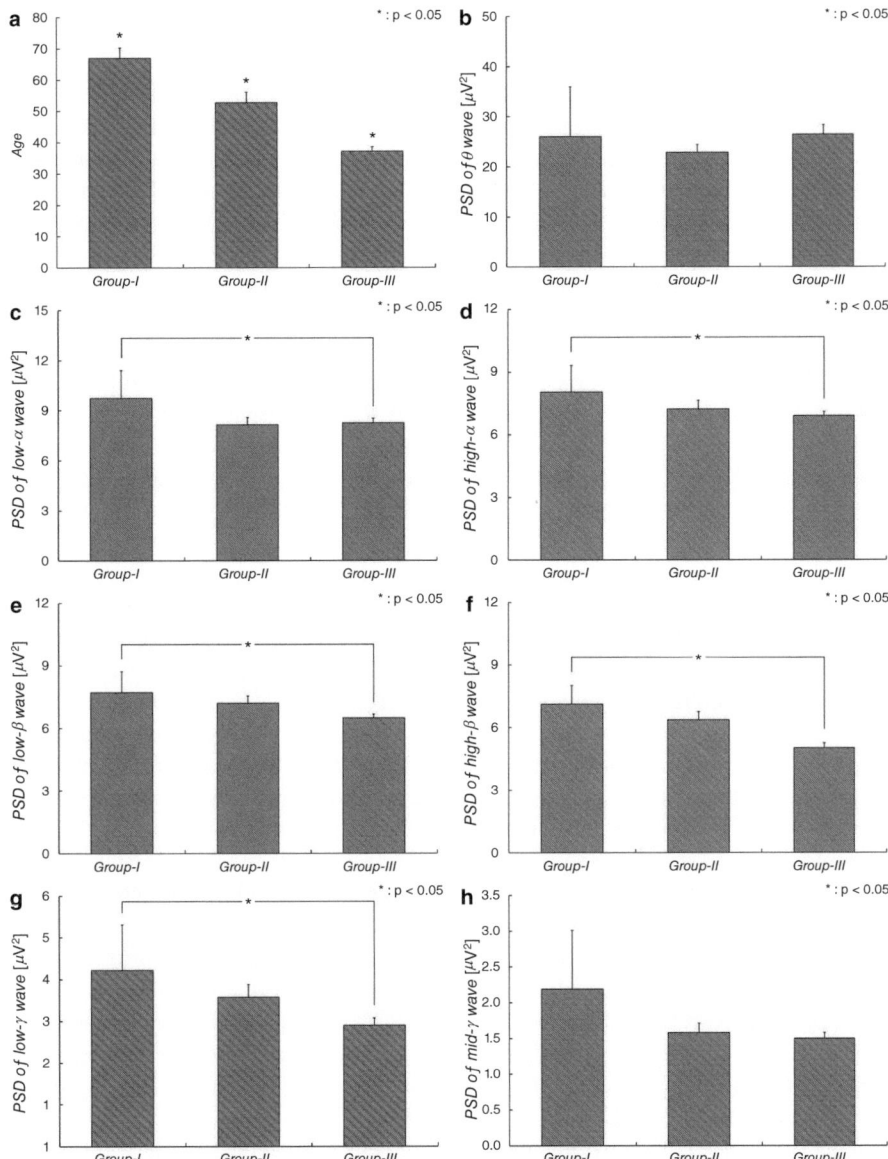

Fig. 5.3 Classification by score (median ± SE): (**a**) age, (**b**) θ, (**c**) low-α, (**d**) high-α, (**e**) low-β, (**f**) high-β, (**g**) low-γ, and (**h**) mid-γ

5.5 Discussion

The visuospatial brain exercise 3D program we are developing is an application of sports vision training. Using this framework, we aim to detect MCI at an early stage by performance evaluation and improve or suppress the progression of symptoms. Based on the mathematical and statistical analysis results of the EEGs measured during undertaking content (3) of the visuospatial brain exercise 3D program, each power spectrum exhibited that the θ wave component was significantly large. It was indicated that θ waves are closely related to the hippocampus, which controls short-term memory [12]. Therefore, content (3) is expected to satisfy the abovementioned objectives.

In this experiment, 160 subjects were classified by age and score groups; accordingly, statistical processing was conducted. In the age group-based classification, those aged 65 years old or more were set as Group A, those aged 50–64 years old were set as Group B, those aged 40–49 years old were set as Group C, and those aged 39 years old or younger were set as Group D. When multiple comparisons were conducted regarding the score differences between groups, we observed significant differences in all comparisons. This is thought to represent the decreased cognitive function associated with aging. When focusing on the integrated PSD value that was calculated based on the EEG results, the low-α, high-α, low-β, high-β, and low-γ values significantly decreased in Group D compared to those in Group A. Meanwhile, there were no significant differences between groups for the θ and mid-γ values. Although the elderly group aged 65 and older exhibited decreased cognitive function, their brain activity was characterized by a different band from that of the young group aged 39 years old and younger. Next, during the score group-based classification, those with a score of 75 points or less were set as Group I, those with a score of 76–85 points were set as Group II, and those with a score of 86 points or more were set as Group III. Subsequently, multiple comparisons were conducted between all groups. Results indicated that there was a significant difference in age between groups in all comparisons. This result also confirms that the subject's age affects the score. When focusing on the integral PSD values calculated based on the EEG results, the low-α, high-α, low-β, high-β, and low-γ values significantly decreased in Group II compared to those in Group I. These results indicated that the relationship between aging and scores may appear in the low-α, high-α, low-β, high-β, and low-γ EEGs.

Finally, we used the abovementioned simple electroencephalography to improve the efficiency of EEG measurements conducted in this experiment. This simple electroencephalograph has already been compared with other medical electroencephalographs, and its measurement accuracy is sufficiently guaranteed [13, 14]. In the future, we will further increase the number of subjects and conduct measurements at nursing homes. Additionally, we will conduct simultaneous measurements using the near-infrared light method (fNIRS); subsequently, brain activity will be measured from multiple angles, and physiological meaning will be assigned to the measured values.

5.6 Future Tasks

The visuospatial brain exercise 3D program we are developing is an application of sports vision training, which is expected to contribute to the early detection of MCI through performance evaluations to either improve them or suppress the progression of symptoms. Based on the mathematical and statistical analysis results of the EEGs measured during work using content (3) of the visuospatial brain exercise 3D program, each power spectrum demonstrated that the θ wave component was significantly large. Accordingly, we determined that the abovementioned purpose was achieved. Furthermore, it was indicated that elderly people aged 65 years and older may have brain activity that is characterized by different bands compared to young people.

Acknowledgments This work was supported by the Descente and Ishimoto Memorial Foundation for the Promotion of Sports Science.

References

1. Statistics Bureau, Ministry of Internal Affairs and Communications, "elderly in Japan from a statistical perspective: 'respect for the aged day'", 2021.
2. Japanese Society of Neurology, "Clinical Practice Guidelines for Dementia" Creation Committee (ed.). Clinical Practice Guidelines for Dementia 2017. Tokyo: Igaku Shoin; 2017.
3. Petersen RC, Smith GE, Waring SC, Ivnik RJ, Tangalos EG, Kokmen E. Mild cognitive impairment: clinical characterization and outcome. Arch Neurol. 1999;56(3):303–8.
4. Suzuki H, Yasunaga M, Naganuma T, Fujiwara Y. Validation of the Japanese version of the Montreal cognitive assessment to evaluate the time course changes of cognitive function: a longitudinal study with mild cognitive impairment and early-stage Alzheimer's disease. Jpn J Geriatr Psychiatry. 2011;22(2):211–8.
5. Nakashima K, Shimohama S, Tomimoto H, et al. Dementia treatment handbook. 2nd ed. Tokyo: Igaku Shoin; 2021.
6. Mendez MF, Cummings JL. Dementia: a clinical approach. 3rd ed. Boston: Butterworth-Heinemann; 2003.
7. Goodale MA, Milner AD. Separate visual pathways for perception and action. Trends Neurosci. 1992;15:20–5.
8. Thiyagesh SN, Farrow TFD, Parks RW, et al. The neural basis of visuospatial perception in Alzheimer's disease and healthy elderly comparison subjects: an fMRI study. Psychiatry Res. 2009;172(2):109–16.
9. Takada H, Miyao M, Takada M, Kinoshita F, Tahara H. Development of sports vision training system using virtual reality for prevention of mild cognitive impairment. Descente Sports Sci. 2019;40:97–109.
10. Kinoshita F, Takada H. A study on the development of VR content for quantitative evaluation of impaired visuospatial ability. In: HCI international 2022–late breaking papers: HCI for health, well-being, universal access and healthy aging: 24th international conference on human-computer interaction, HCII 2022, virtual event, June 26–July 1, 2022, proceedings. Cham: Springer Nature; 2022. p. 440–50.

11. Oohara T, Kawaguchi K, Takeuchi T, Kinoshita F. Visuospatial ability of elderly using head-mounted display with gaze measurement function. In: HCI international 2023: universal access in human-computer interaction: 25th international conference on human-computer interaction, HCII 2023, virtual event, July 23–July 28, 2022, proceedings. Cham: Springer Nature; 2023. p. 81–90.
12. Pfurtscheller G, Stancák A Jr, Neuperb C. Event-related synchronization (ERS) in the alpha band—an electrophysiological correlate of cortical idling: a review. Int J Psychophysiol. 1996;24(1–2):39–46.
13. Ratti E, Waninger S, Berka C, Ruffini G, Verma A. Comparison of medical and consumer wireless EEG Systems for use in clinical trials. Front Hum Neurosci. 2017;11:398.
14. Mendez GR, Dunwell I, Mirón EAM, Cerdán MDV, de Freitas S, Liarokapis F, Gaona ARG. Assessing NeuroSky's Usability to Detect Attention Levels in an Assessment Exercise. In: Lecture Notes in Computer Science (LNCS), vol. 5610; 2009. p. 149–58.

Chapter 6
Comparison of Radial Motions in the Young with Those in the Elderly While Viewing 3D Video Clips Using Artificial Intelligence

Yoshiki Itatu, Hiroki Takada, and Tomoki Shiozawa

Abstract It has been reported that motion sickness is not visually induced by the foveal vision but by the peripheral vision. As a side note, different findings in the radial motion were observed for each vision. The aim of this study is to classify two types of radial motion data measured while viewing 3D video clips during the foveal and the peripheral vision, respectively. Healthy subjects, composed of ten young males and six elderly persons, voluntarily participated in this study. We herein extracted about 2300 time series of these radial motions for each data size to tell the foveal vision from the peripheral vision by statistical machine learning. As a result, statistical significance was observed in the accuracy of the classification, which increased with the data size. The maximum value of the accuracy was obtained, where the data size was set to be 10 s.

Keywords Visually induced motion sickness (VIMS) · Artificial intelligence (AI) · Statistical machine learning (ML) · Radial motions · Flick · Fixation micro-movements · Peripheral vision · Foveal vision

---

Y. Itatu
Graduate School of Engineering, University of Fukui, Fukui, Japan

H. Takada (✉)
Department of Human and Artificial Intelligence Systems, Graduate School of Engineering, University of Fukui, Fukui, Fukui, Japan
e-mail: takada@g.u-fukui.ac.jp

T. Shiozawa
Department of Business Administration, Aoyama Gakuin University School of Business, Tokyo, Japan

6.1 Introduction

In recent years, we have had more and more opportunities to experience stereoscopic films in our daily lives, which is caused by the development of video technology. In addition, research and development of stereoscopic imaging technology is active in the fields of education and medicine [1, 2]. Various 3D video display systems such as mobile devices, free-viewpoint TV, and 3D cinema have been developed. A few recent displays can also present binocular and multi-aspect autostereoscopic images, although 3D glasses are generally required. However, in either case, there are issues with unpleasant symptoms such as headaches, vomiting, eye strain, and a lack of ambience and realism in the images [3, 4]. Especially in Japanese 3D televisions, dynamic movements cannot be fully expressed since the binocular disparity is set to one degree or less. Excessive measures against visually induced motion sickness (VIMS) have been implemented without an appropriate manufacturing standard for stereoscopic video clips (VPs) and their display systems since the eye strain induced by 3D video viewing has not yet been still elucidated [5, 6].

In 3D video viewing, it is generally understood that the lens is accommodated to the depth of the screen that displays the image, whereas the eyes converge at the position of the 3D object, which is a common idea of eye strain with 3D image viewing. As might be expected, the convergence and lens accommodation are consistent in natural vision. The discrepancy is considered to be the cause of the eye strain and the motion sickness induced by stereoscopic viewing [7, 8]. According to the sensory discrepancy theory, VIMS is caused by another process. We usually acquire sensory information about our position to keep our balance and movement control. However, there is a discrepancy with the sensory information we have acquired in our daily lives when we are placed in a new dynamic environment, such as riding in a car or on a boat. At this moment, the neural network is recombined in our brain to adopt new sensory information. In the process of developing motion sickness, the disturbance of spatial orientation is noticed somewhere in the brain [9]. It has been reported that there are significant differences in the severity of VIMS among visual recognition methods, although the cause of VIMS has not been elucidated [10].

There is a difference between the radial motions during foveal or peripheral vision, in which human visual information can be classified and processed. Rod cells are involved in sensing visual information in the peripheral visual field and play a role in light/dark discrimination, whereas pyramidal cells (PCs) are densely distributed in the macula of retina. The PCs contribute to high-definition vision in the foveal visual field (CVF) [11] and the function of sensing colors. In the brain, visual information is processed in the ventral visual tract.

Rod cells (RCs) are responsible for peripheral and night vision. Compared to the PCs, the RCs are not able to perceive colors and/or to make images in the CVF clear. However, there are a number of cells and the RCs are more sensitive to light [12–15]. The RCs have a lower spatial resolution because they are sparsely distributed mainly in the retina's periphery and are not present in the fovea foveaIis, the area with the highest resolution. Visual information is processed in the brain in the dorsal visual tract [16]. The dorsal stream extends from the most caudal part of the occipital lobe to the parietal lobe, which is known to be essential for grasping three-dimensional relationships among objects in space. It is pointed out that the function at the beginning of the dorsal stream differs from that at the end. The former is the visual function, and the latter is the spatial perception.

Our experimental studies have shown that peripheral vision deteriorates with advancing age [17]. The decline of peripheral vision can also be seen in traffic accidents involving elderly drivers. The parietal lobes in the elderly would not receive the information obtained from peripheral vision. In addition, a decrease in the blood flow is prominent in the parietal cortex of mild cognitive disease (MCI), especially in Alzheimer's disease [18]. Therefore, we believe that the elderly, whose peripheral vision function is impaired, a method of viewing that tends to induce VIMS, may not be able to view 3D images with peripheral vision compared to the young.

It is said that the majority of human cognition is based on information obtained from the vision [19]. That is why the human intellectual process may be involuntarily reflected in the radial motion. Therefore, understanding the general characteristics of the radial motion would provide a clue to the elucidation of human information processing.

In this paper, we measure the radial motion of the elderly while viewing 3D video clips in the periphery or foveal vision. Except for the former, subjects followed a sphere as a visual target on a stereoscopic video clip in accordance with the following protocol. Radial motion is expected to be controlled in the foveal vision while tracking the target. That is why this radial motion is herein assumed to be controlled in the foveal vision. Data sequences for several sample sizes were extracted from time series data of the radial motion measured in this research. These time series were used as a data set, and deep learning was conducted to classify the data into two types: data sequences for the peripheral vision and those for the foveal radial motion data.

6.2 Material and Method

The contents of the experiment were fully explained to these subjects before the experiment. Their written consent was obtained. In addition, this study was conducted with the approval of the Research Ethics Committee for Human Subjects, Department of Human and Artificial Intelligent Systems, Graduate School of Engineering, University of Fukui (H2019003).

The experiment was conducted in a dark room, and the subjects stood with the Romberg posture [20] for 60 s of each measurement. The radial motion was recorded while viewing the following video clips that were reconstructed from Sky Crystal with the company's permission (Olympus Memory Works, Tokyo). The size of the sphere of the presented image was 6 [deg] for near and 0.5 [deg] for far. Also, the viewing angle of the background was set to be 23 [deg]. In this experiment, a 55-in. 3D display 55UF8500 (LG, Soul) was used to view the video clips in either peripheral or foveal vision in which subjects keep tracking on spherical visual targets. The viewing distance was set at 2 [m] in this experiment. The radial motion was measured using an eye mark recorder, the EMR-9 (Nac Image Technology, Tokyo). The position of the gaze was recorded at each sampling time while viewing the video clips. The sampling frequency was set to be 60 [Hz]. We recorded xy coordinates [pix] of the viewpoint at each sampling time while viewing video clips. Time series data for horizontal/vertical components of the LCD screen were collected in the x/y direction, where the right/upper directions were set to be positive, respectively.

6.2.1 Experiment 1

The experiment was conducted on ten healthy young subjects (22.67 ± 0.78 years). The experiment for the young was designed as a study for stereoscopic viewing. In this experiment, the radial motions of the young were measured while viewing 3D video clips without a constriction in the background (Fig. 6.1a).

6.2.2 Experiment 2

The experiment was conducted on six healthy elderly subjects (75.0 ± 8.2 years). The experiment for the elderly was designed as a study for tunnel vision [21]. The video clips were herein composed of 3D images with and without a constriction in the background (Fig. 6.1).

6.2.3 Machine Learning

The performance of a machine learning model for the classification can be quantified and evaluated by obtaining a confusion matrix, which is a square matrix consisting of four components: True Negative (TN), False Positive (FP), False Negative (FN), and True Positive (TP). The correct rate, precision, and recall

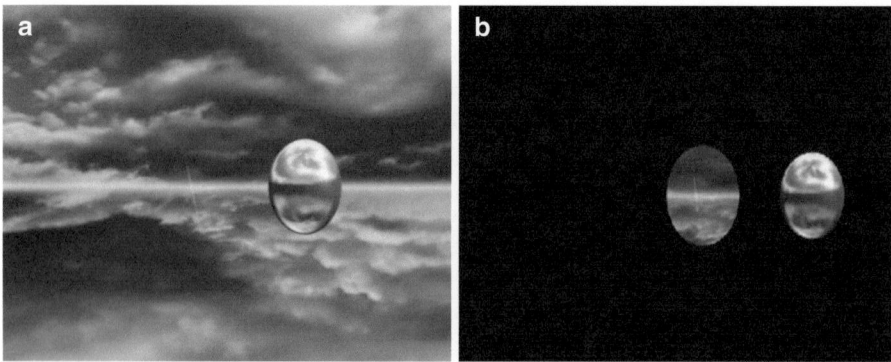

Fig. 6.1 Images included in the video clips. A sphere is moving smoothly in 3D space (**a**). A black background frame is fixed to each image (**a**) to weaken the influence of the surroundings (**b**)

resulted from the predicted results of all test data as TN + TP, TP + FP, and TP + FN, respectively.

We extracted data sequences of the radial motion with the sample size of 1, 2, 3, 4, 5, 6, 7, 8, 9, 10, and 20 s for the foveal and the peripheral vision, respectively (Table 6.1). Comparing accuracy for the classification between the radial motion in the peripheral and that in the foveal vision, eleven kinds of these sample sizes were set to use the optimal short period of time, which allows for real-time analysis and systems. These datasets were used for statistical machine learning to classify any test data into the peripheral and foveal vision [22]. The dataset used for the training was labeled with the peripheral vision as zero and the foveal vision as +1. The leave-one-out method was used as the validation method for the training and test data [23]. For the training model, we used a model that combines the Gated Recurrent Unit (GRU), a gating mechanism in regression neural networks [24], and the CNN, which can capture spatial information [25] (Fig. 6.2).

Table 6.1 The number of data sequences for each sample size

Sample size of time series (s)	Data sequences
1	2380
2	2340
3	2300
4	2260
5	2220
6	2180
7	2140
8	2100
9	2060
10	2020
20	1620

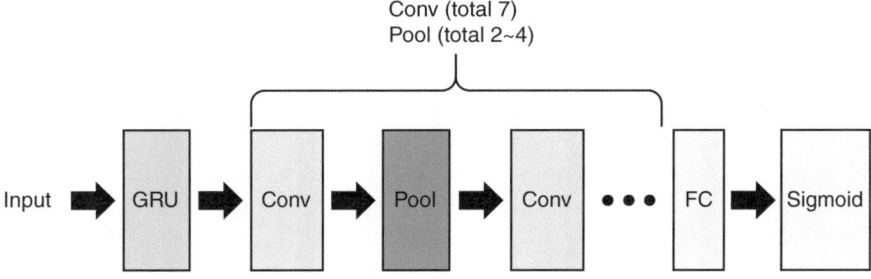

Fig. 6.2 Model structure

6.3 Result

6.3.1 Accuracy

Values of the correct rate among the sample sizes were calculated for the young and the elderly (Fig. 6.3). The Friedman tests were conducted at a significance level of 0.05 to compare the correct rate among the sample sizes. Since significant differences were found among the sample sizes for each group of the young and the elderly, multiple comparisons were also conducted for every combination of the sample sizes by Scheffe's method. The results showed significant differences in the correct rate for the young among six pairs of the sample sizes shown in Fig. 6.3 ($p < 0.05$).

Comparing the correct rate of the elderly with that of the young, the latter tended to be greater than the former. In addition, the correct rate increased with the sample size for both groups. The maximum value of the correct rate for the elderly was obtained as 89.0% at the sample size of 20 s, whereas that for the young was 97.2% at 10 s.

6.3.2 Precision and Recall

In Figs. 6.4 and 6.5, we compared the values of precision and recall for each sample size (Figs. 6.4 and 6.5). According to the Receiver operating characteristics (ROC) analysis [26, 27], the recall rate tended to be higher than the precision. This

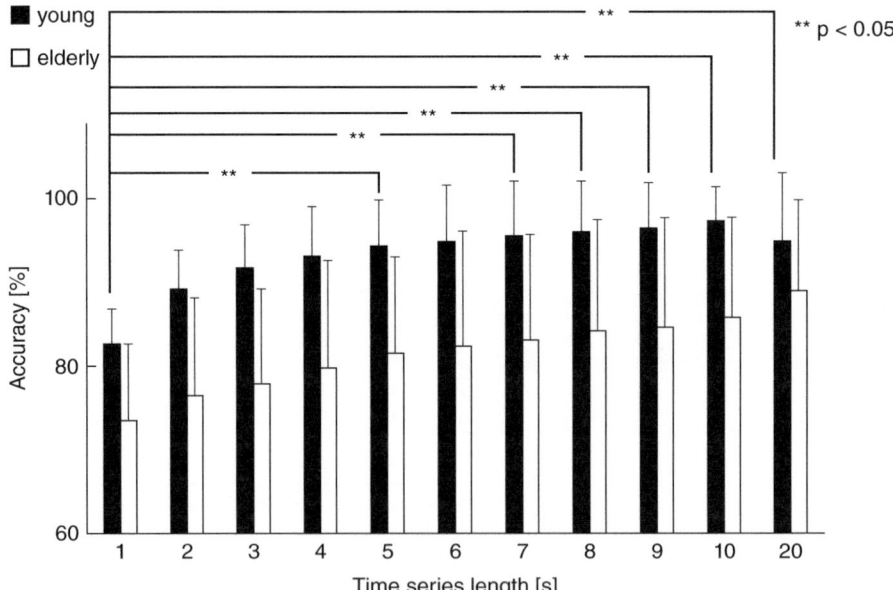

Fig. 6.3 Correct rate of the classification in the young and elderly for each sample size

6 Comparison of Radial Motions in the Young with Those in the Elderly While... 83

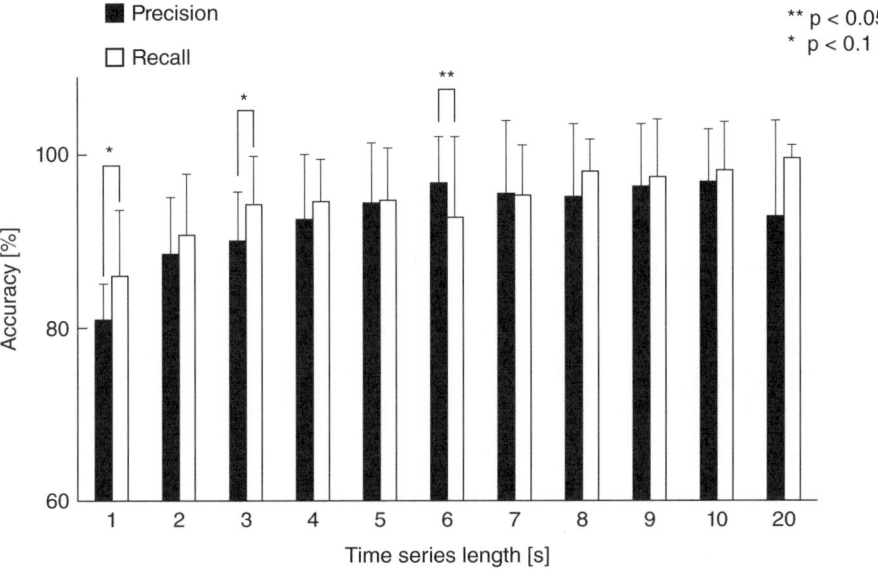

Fig. 6.4 Correct rate of the classification in the young for each sample size

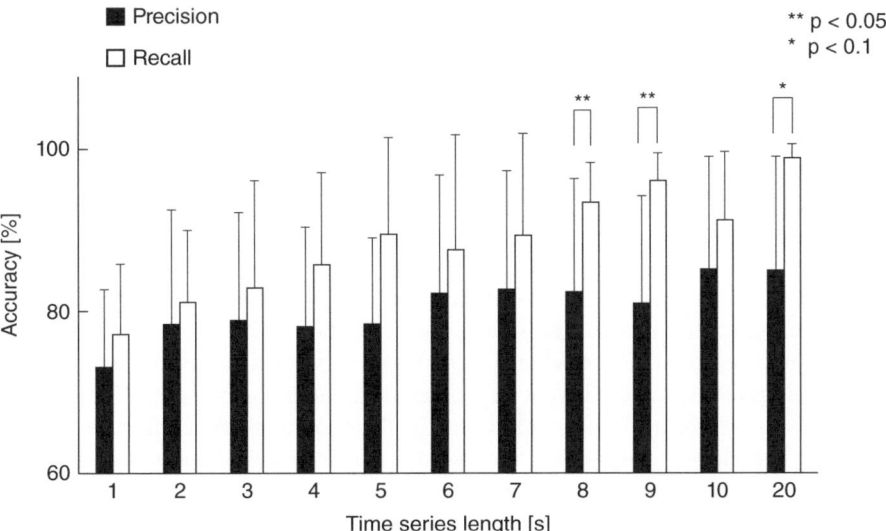

Fig. 6.5 Correct rate of the classification in the elderly for each sample size

indicates that values of the correct classification rate for the foveal vision are larger than those for peripheral vision.

For the value of the correct rate, the statistical test was conducted to test whether there was a difference in the correct rates among the sample size at the 5% level of significance. The results showed that there was a significant difference in the sample size. In addition, the Wilcoxon signed-rank test was also performed to test whether there was a difference in the accuracy between the precision and the recall rate at the 5% level of significance. A significant difference between precision and recall values was observed from the elderly's data of the sample size 8–9 s ($p < 0.05$). Also, statistical differences tended to be observed between the precision and the recall rate for 1 s and between those for 20 s ($p < 0.1$). In the case of the young, significant difference between precision and recall values was observed from the sample size 6 s ($p < 0.05$). Also, statistical differences tended to be observed between the precision and the recall rate for 1 s and between those for 3 s ($p < 0.1$).

6.4 Discussion

6.4.1 Accuracy

As a result, the accuracy values in the young tended to be higher than those in the elderly. The elderly were asked to perform peripheral vision; however, they may have unconsciously used foveal vision to obtain visual information and maintain their equilibrium. Also, the properties of the elderly could not be extracted in the radial motion well because the number of subjects in the elderly was smaller than that in the young. Furthermore, the age range of the elderly was wider than that of the young. As in Fig. 6.3, values of the standard deviation were larger for the elderly compared to those for the young.

6.4.2 Precision and Recall

The precision and recall rate results for each sample size indicate that the values of the recall rate are greater than those of the precision. This indicates that values of the correct classification rate for the foveal vision are larger than those for peripheral vision. We considered that the elderly person was instructed to look at the periphery but ended up looking at the foveal due to deterioration in the peripheral vision. The human control system would be influenced by the poor in this function with advancing age. The elderly cannot help but use the foveal vision to keep their standing posture in the gravitational environment.

The recall rate increased, but the precision decreased after the sample size increased from 10 s to 20 s. This suggests that the peripheral vision has been classified as foveal vision because the involuntary saccadic movements that occurred several times per 20 s during peripheral vision were similar to the flicks included in the fixation micro-movements during the foveal vision.

6.5 Conclusion

The aim of this study was to measure the radial motion while the subjects viewed 3D video clips in the peripheral or foveal vision. Using deep learning, we have succeeded in classifying these data. Ten healthy young males and six healthy elderly subjects voluntarily participated in this study to measure the radial motion while viewing the video clips in which a sphere smoothly moved between the proximal and distal sides of the image.

Comparing values of the correct rate in the elderly with that in the young for each sample size, the latter tended to be higher than the former (Fig. 6.3). In addition, values of the correct rate increased with the size of the series for both the elderly and the young. However, in the case of the young, the maximum value of the correct rate was obtained when the sample size was set to 10 s.

Comparing values of the precision with those of the recall for each sample size, the recall tended to be greater than the precision for both the elderly and the young group. This indicates that the accuracy of the classification involved in the foveal vision was higher than that in the peripheral vision for all sample sizes (Figs. 6.4 and 6.5). Furthermore, these values tended to increase proportionally to the sample size for both groups. Although the recall rate increased, the precision decreased after the sample size was increased from 10 s to 20 s. This suggests that the peripheral vision has been classified as the foveal vision because the involuntary saccadic movements that occurred several times per 20 s during peripheral vision were similar to the flicks included in the fixation micro-movements during the foveal vision.

Also, these values of accuracy for data classification with AI tended to be higher for the young than for the elderly. This tendency was obtained for each sample size. As a side note, the standard deviation tended to be higher for the elderly than for the young. In this study, we might not have succeeded in extracting the features from the radial motions of the elderly. As discussed at the beginning of this paper, the elderly may not be able to view 3D images in their peripheral vision. As a result, some of the elderly were able to perform peripheral vision while others were not.

In this study, two types of data, radial motions in the peripheral and the foveal vision, were classified for each sample size of the young and the elderly data. As a result, values of the correct rate increased in proportion to the sample size for both the elderly and the younger. In the next step, a comprehensive analysis of the sample size is expected. Using the same kind of the AI, components of the radial motions might be extracted as flick (microsaccade), drift, and tremor [28]. According to the distribution of their components, dependency on the sample size to classify the data might be clarified to propose the optimal measurement time for the radial motion.

References

1. Setozaki T, Morita Y, Takeda T. Development of multi aspect type VR teaching material based on needs investigation and practice class. Trans Virtual Real Soc Japan. 2006;11(4): 537–43.
2. Kubota Y, Yamasita J, Okumura S, Kuzuoka H, Hiroshi H. Learning and evaluation the moon phases using a solar system simulation. J Sci Educ Japan. 2007;31(4):248–56.
3. Allen RC, Singer MJ, McDonald DP, Cotton JE. Age differences in a virtual reality entertainment environment, a field study. Proc Hum Factors Ergon Soc Annu Meet. 2000;44(5): 542–5.
4. Hamagishi G. Ergonomics fbr3D displays and their standardization. Inst Image Inform Television Eng. 2009;33(16):9–12.
5. Yano S. Size of disparity for binocular fusion—a study on stimulus target properties. Trans Inst Electron Inform Commun Eng. 1991;75(10):1720–8.
6. Emoto M, Yano S, Nagata S. Distribution of fusional vergence limit in viewing stereoscopic image systems. J Inst Image Inform Television Eng. 2001;55(5):703–10.
7. Toates FM. Vergence eye movements. Doc Ophthalmol. 1974;37(1):153–214.
8. Hoffman HG, Patterson DR, Seibel E, Soltani M, Jewett-Leahy L, Sharar SR. Virtual reality pain control during burn wound debridement in the hydrotank. Clin J Pain. 2008;24: 299–304.
9. Uno A, Nakagawa A, Hori A, Takada N, Kubo T. Neural substrate for motion sickness: involvement of the limbic system. Equilibrium Res. 2006;65(4):213–22. (In Japanese).
10. Takada M, Fukui Y, Matsuura Y, Sato M, Takada H. Peripheral viewing during exposure to a 2D/3D video clip, effects on the human body. Environ Health Prev Med. 2015;20(2):79–89.
11. Tachibanaki S, Kawamura S. Molecular bases of the difference between rod- and cone-mediated vision. Jpn Soc Comp Physiol Biochem. 2017;34(3):70–9.
12. Fu Y, Yau KW. Phototransduction in mouse rods and cones. Pflugers Archiv Eur J Physiol. 2007;454(5):805–19.
13. Kawamura S, Tachibanaki S. Rod and cone photoreceptors: molecular basis of the difference in their physiology. Comp Biochem Physiol A Mol Integr Physiol. 2008;150(4):369–77.
14. Yau KW, Hardie RC. Phototransduction motifs and variations. Cell. 2009;139(2):246–64.
15. Rieke F, Baylor DA. Single-photon detection by rod cells of the retina. Rev Mod Phys. 1998;70(3):1027–36.
16. Zeki SM. Functional organization of a visual area in the posterior bank of the superior temporal sulcus of the rhesus monkey. J Physiol. 1974;236(3):549–73.
17. Ono R, Takada H. Comparison of eye movement while viewing a 3D video clips among ages. Inst Electr Eng Japan. 2021;141(6):752–3.
18. Matsuda H. Neuroimaging of Alzheimer's disease. Japan Geriatr Soc. 2012;49(4):425–30.
19. Zimmermann M. Neurophysiology of nociception. In: Poter R, editor. International review of physiology. Neurophysiology II, vol. 10. Baltimore: University Park Press; 1976. p. 179–221.
20. Ben-David J, Podoshin L, Fradis M. A comparative Cranio-Corpography study on the findings in the Romberg standing test versus the Unterberger/Fukuda stepping test in vertigo patients. Acta Otorhinolaryngol Belg. 1985;39(6):924–32.
21. Itatsu Y, Sugiura A, Takada H. A Study for Statistical Machine Learning to Classify Radial Motions of the Elderly. In: Proceedings of the 2021 16th International Conference on Computer Science & Education (ICCSE); 2021. p. 815–8. https://doi.org/10.1109/ICCSE51940.2021.9569478.
22. Itatsu Y, Matsuura Y, Shiozawa T, Takada H. A study for statistical machine learning to classify radial motions. FORMA. 2022;37(2):S17–22.
23. Stone M. Cross-validation and multinomial prediction. Biometrika. 1974;61(3):509–15.

24. Kyunghyun C, Bart VM, Caglar G, Dzmitry B, Fethi B, Holger S, Yoshua B. Learning phrase representations using RNN encoder-decoder for statistical machine translation. In: Proceedings of the 2014 conference on empirical methods in natural language processing (EMNLP); 2014. p. 1724–34.
25. Wang ZJ, Turko R, Shaikh O, Park H, Das N, Hohman F, Kahng M, Chau DH. CNN Explainer: learning convolutional neural networks with interactive visualization. In: arXiv preprint arXiv: 2004.15004; 2020. p. 1–11.
26. Tom F. An introduction to ROC analysis. Pattern Recogn Lett. 2006;27(8):861–74.
27. Olson DL, Delen D. Performance evaluation for predictive modeling. In: Olson D, Delen D, editors. Advanced data mining techniques. Heidelberg: Springer; 2008. p. 137–47.
28. Pritchard RM. Stabilized images on the retina. Sci Am. 1961;204(6):72–8.

Part II
Aging Society and Gerontology

Chapter 7
Elderly Migration in Tokyo Metropolitan Area, Japan

Makoto Hirai

Abstract As the largest metropolitan area in Japan, the Tokyo Metropolitan Area (TMA) was one of the earliest areas to observe an elderly migration, and both the rate and volume of migration continue to surpass the national average, as will be described in detail later. The TMA is, therefore, the area with the most active elderly migration in Japan. This chapter describes the characteristics of elderly migration in the TMA. Elderly migration in the TMA is characterized by a comparatively high rate of inter-prefectural migration than in other areas and greatly impacts the distribution of the elderly population in Japan as a whole. Inter-prefectural migration in the TMA is high for both in- and out-migration, but trends change according to migrant age. Compared to the young-old migrants who migrated more often from the TMA to non-metropolitan areas such as the Tohoku or Kyushu regions, the reverse movement of non-metropolitan areas to the TMA was notable among old-old migrants. This is thought to be related to the characteristics of the TMA that have traditionally drawn in a large population and are the very characteristics of elderly migration in the TMA.

Keywords Elderly population · Elderly migration · Migration preference index · Tokyo Metropolitan Area

M. Hirai (✉)
Kanagawa University, Yokohama, Kanagawa, Japan
e-mail: m-hirai@kanagawa-u.ac.jp

7.1 Introduction

Japan is one of the oldest countries in the world, with 28% of the population 65 years and older in 2020. Among these, the Tokyo Metropolitan Area (TMA), the largest metropolitan area in the nation (defined as the capital Tokyo and Chiba, Saitama, and Kanagawa prefectures in this study) has over 9.05 million people aged 65 years old as of 2020, which accounts for approximately one-fourth of older adults in Japan. The TMA is the area with the highest density of older adults in Japan, and the demographic changes of this large population greatly impact various aspects of society, such as healthcare, welfare, economics, and local communities. The distribution and dynamics of older adults in the TMA are essential focus areas in discussions on the effects of aging on a population.

Changes in the size of the elderly population of a region are affected not only by natural increases due to the aging of residents and decreases due to their deaths but also by social increases and decreases such as inflows and outflows due to the movement of older adults [1]. Therefore, focusing on elderly migration and collecting trend data is an essential challenge in forecasting changes in a community.

An increasing number of studies have researched elderly migration in Japan since Otomo (1981) [2] first examined the topic through an analysis of the 1970 Population Census and found an increase in the migration rate of older adults in metropolitan areas. In an analysis of migration data from the 1980 Population Census, Uchino (1987) [3] characterized elderly migration as "generally higher in areas undergoing notable urbanization, and lower in rural prefectures." This trend has been confirmed by Tahara and Iwadare [4] in their analysis of migration data in the 1990 Population Census. As the largest metropolitan area in Japan, the TMA was one of the earliest areas to observe an elderly migration, and both the rate and volume of migration continue to surpass the national average, as will be described in detail later. The TMA is, therefore, the area with the most active elderly migration in Japan.

This chapter describes the characteristics of elderly migration in the TMA. In the next section, the characteristics of the elderly migration of the TMA will be examined with reference to the trends in the nation as a whole. Section 7.3 will focus on the people choosing migration. Finally, Sects. 7.4 and 7.5 will highlight interprefectural migration, the subtype of elderly migration in the TMA with the greatest influence on the regional distribution of older adults, to analyze the local characteristics of migration rates and trends in migration.

The data used in this chapter is the population migration statistics of the 2020 Population Census. The Population Census defines migration as having a different address at the time of the survey from the one recorded as the usual residence 5 years earlier. Demographic details of the migrant, such as age, gender, and occupation, are also tallied and published in the Population Census.[1] Within this

---

[1] The Population Census defines migration as a difference between the usual residence 5 years ago and the usual residence at the time of the survey. Therefore, it is not possible to capture all movements within the survey period, such as people who moved multiple times within the 5-year survey period.

chapter, "elderly migration" is defined as the migration of people "65 years and older." Subsequently, this chapter focuses on migrating people aged 60 and older as of 2015. Unless otherwise noted, the age categories used in the text, figures, and tables of this chapter are made with reference to the age at the end of the fiscal year 2020.

7.2 Outline of Elderly Migration in the Tokyo Metropolitan Area

Elderly migration in the TMA will be first outlined. Table 7.1 illustrates the population of older adults and the migration in the TMA and the nation.[2] Approximately 3.17 million older adults migrated in 2020; their migration rate was 8.2% in Japan. Elderly migration in Japan became visible in the 1980s, and at the time, there were approximately 1.55 million elderly migrants and a 9.5% migration rate (1990 Population Census). As the total elderly population increased drastically in the last 30 years, the migration rate decreased slightly, but the absolute number of migrants doubled in that time.

The total number of migrants in the TMA was 0.8 million, and the migration rate was 8.2%. The migration rate was approximately equal to the national figure, but migrants in the TMA accounted for approximately 25% of all migrants in Japan. Now turning our eyes to the scale of migration, trends of migration in the TMA demonstrate that the migration rate within one's municipality is lower than that of the whole country, but the inter-prefectural migration rate was higher. This suggests that the elderly are migrating across a wider area and that the TMA is a major component of the change in the distribution of the elderly population in Japan. The frequency of inter-prefectural migration characterizes elderly migration in the TMA.

The Population Census does not provide data on inter-national out-migration, which has been drawing attention as a destination of post-retirement migration; thus, the data on inter-national migration is limited to that of inbound migration. The number of elderly migrants from abroad is 11,000 nationwide, which is extremely low as a percentage of the elderly population. Therefore, this chapter focused only on migration within Japan for analysis.

---

[2] Migration statistics in the Population Census have a long observation period of 5 years. In the case of older adults, a great fluctuation due to a decrease in deaths can occur in 5 years. Therefore, various migration rates in this chapter were calculated as a ratio of the migrations to the population's average at the beginning of the period (population aged 60 years and over in 2015) and the end of the period (population aged 65 and over in 2020).

Table 7.1 Elderly migrants in the Tokyo Metropolitan Area and Japan (2020)

Scale of migration	Tokyo Metropolitan Area		Japan	
	Number of migrants (people)	Migration rates (%)	Number of migrants (people)	Migration rates (%)
In same municipalities	498,466	5.1	2,164,121	5.6
Inter-municipalities (in same prefecture)	149,396	1.5	588,223	1.5
Inter-prefecture	151,576	1.5	402,716	1.0
International	4984	0.1	11,407	0.0
Total	804,422	8.2	3,166,467	8.2

(Source: Population Census)

7.3 Characteristics of Elderly Migrants in the Tokyo Metropolitan Area

This chapter outlines the characteristics of elderly migrants in the TMA. First, it has been noted that most elderly migrants are females [5]. In 2020, the number of elderly migrants in the TMA was 391,000 males and 579,000 females, with a sex ratio of 67.6 (Table 7.2). This trend is observed regardless of the scale of migration.

Regarding the number of migrants by age, the number of migrants among relatively young-old men aged 65–69 and 70–74 years is about the same as that of women. However, the number of migrants aged 85 years and older falls to approximately half the number of migrants aged 65–69 years. Conversely, the number of migrations increases with age in women. Therefore, the sex difference in migrants narrows with age. Elderly migration in the TMA should not be regarded as a phenomenon peculiar to women, but attention should also be paid to men and women (couples) in early old age.

Concerning the number of inter-prefectural migrations, out-migration exceeds in-migration for both males and females aged 65–69 and 70–74 years, whereas in-migration exceeds out-migration in women 75 years and older, and in men as well at 85 years and older. That is, the direction of migration reverses with age.

The migration of older adults is also called "retirement migration"; thus, the employment status of migrants will be observed. Of the TMA elderly migrants, the rate of those who continued to work after the migration (2020) was higher in the men aged 65–69 years (e.g., 56.6% of within-municipality migrants and 51.3% of same-prefecture inter-municipality migrants). Such a relatively close-range migration was presumably possible without leaving an employer. However, only around 50% of close-range migrants were working, and the rate of workers decreases dramatically with age, even in migrations within municipalities with a high rate of workers to 39.9% among 70- to 74-year-old men, and further down to 22.4% among 75- to 79-year-old men. The proportion of inter-prefectural out-migrants working is also low (65–69 years, 42.9%; 70–74 years, 27.6%). Most migrations are assumed to have occurred after retirement. The rate of working women is lower than that of men, where the working rate falls below 30% at 70 years old at the time of

7 Elderly Migration in Tokyo Metropolitan Area, Japan

Table 7.2 Elderly migrants by age and sex in TMA (2015–2020)

	Scale of migration	Age group					Total
		65–69	70–74	75–79	80–84	85+	
Male	In same municipalities	49,154	49,961	34,871	26,531	33,861	194,378
	Inter-municipalities (in same prefecture)	16,032	14,773	10,455	8108	10,629	59,997
	Inter-prefecture (in-migration)	17,908	14,818	9428	7282	9470	58,906
	Inter-prefecture (out-migration)	28,490	21,411	11,687	7933	8855	78,376
	Total	111,584	100,963	66,441	49,854	62,815	391,657
Female	In same municipalities	45,789	50,914	46,719	48,730	111,936	304,088
	Inter-municipalities (in same prefecture)	14,994	15,652	13,685	14,045	31,023	89,399
	Inter-prefecture (in-migration)	16,762	16,793	14,188	14,564	30,363	92,670
	Inter-prefecture (out-migration)	20,889	18,608	14,063	13,383	26,169	93,112
	Total	98,434	101,967	88,655	90,722	199,491	579,269
Sex ratio	In same municipalities	107.3	98.1	74.6	54.4	30.3	63.9
	Inter-municipalities (in same prefecture)	106.9	94.4	76.4	57.7	34.3	67.1
	Inter-prefecture (in-migration)	106.8	88.2	66.5	50.0	31.2	63.6
	Inter-prefecture (out-migration)	136.4	115.1	83.1	59.3	33.8	84.2
	Total	113.4	99.0	74.9	55.0	31.5	67.6

(Source: Population Census)

migration, irrespective of the scale of migration. That is, most elderly migrants in the TMA are non-workers or retirees.

A given fraction of the migration of older adults can be presumed to involve a move into long-term care or another institutional household. In the case of institutional household admission, the location of the institutional welfare household and the waiting list for admission are also factors, which means that the age and scale of migration are also affected.[3] On the other hand, a move into a general household instead of an institutional household will have a large impact on the community, such as consumer behavior and participation in the community and enhancement of home care. Therefore, the residence after the migration will be investigated. The percentage of TMA elderly migrants whose destination was an institutional household is shown in Table 7.3. Regardless of the scale of migration, the percentage of institutional household residents remains low for both genders aged 65–69 and 70–74 years, but the percentage increases with age. This trend is particularly

---

[3] Due to the shortage of special elderly nursing homes in central Tokyo, people are migrating from the 23 wards to institutional households in Okutama-machi, on the western edge of Tokyo. Therefore, the inflow rate of the old-old to Okutama-machi is marked [12].

Table 7.3 Institutional household residents as a percentage of elderly migrants in TMA (2020)

	Scale of migration	Age group				
		65–69	70–74	75–79	80–84	85+
Male	In same municipalities	7.4	11.8	22.3	40.6	68.4
	Inter-municipalities (in same prefecture)	12.5	18.9	31.6	47.3	67.7
	Inter-prefecture (in-migration)	6.7	10.6	16.9	26.3	42.5
Female	In same municipalities	4.5	10.1	25.8	50.0	79.8
	Inter-municipalities (in same prefecture)	7.1	13.9	29.3	50.9	75.6
	Inter-prefecture (in-migration)	2.6	6.0	15.5	30.2	51.0

(Source: Population Census)

noticeable in short-distance migrations (migration within the same municipality), and about 70% of migrations of people aged 85 years and over are admissions to institutional households. However, the admission rate to institutional households is relatively low among inter-prefectural in-migration, and the rates are only 42.5% for men and 51.0% even among migrants 85 years and older. That is, migration of relatively young old adults and inter-prefectural migrations involve a move into a general household rather than an admission to an institutional welfare household.

Thus, elderly migration in the TMA mainly involves retirees, and there is not much gender difference between young and old migrants, but the migration of women exceeds that of men as they age. Regarding the scale of migration, inter-prefectural migration comprises a relatively large proportion of moves. Inter-prefectural migrants have a relatively low admissions rate to institutional households; most move into general households after the migration. In addition, the relatively young were characterized by an excess of out-migration, but an excess of in-migration was observed with age, indicating that the direction of migration was reversed. Therefore, in the following sections, we will analyze inter-prefectural migration among elderly migration in the TMA.

7.4 Characteristics of Inter-Prefectural Elderly Migration of the Tokyo Metropolitan Area

Figure 7.1 presents the rates of elderly inter-prefectural migration according to the 2020 Population Census by prefecture (Fig. 7.1). In the figure, the straight line connecting the bottom left (0, 0) and top right (2.5, 2.5) connects the points where the in-migration and out-migration rates are equal. Points to the bottom right of this line represent greater in-migration rates, and points to the top left of this line represent greater out-migration rates. Of the 47 prefectures, 29, representing over half of the prefectures, have greater in-migration rates. The lowest in-migration rate was observed in Hokkaido (0.47%), and the highest was observed in Saitama (1.76%), and the lowest out-migration rate was also observed in Hokkaido (0.45%). The highest out-migration rate was observed in Tokyo (2.48%). There were active

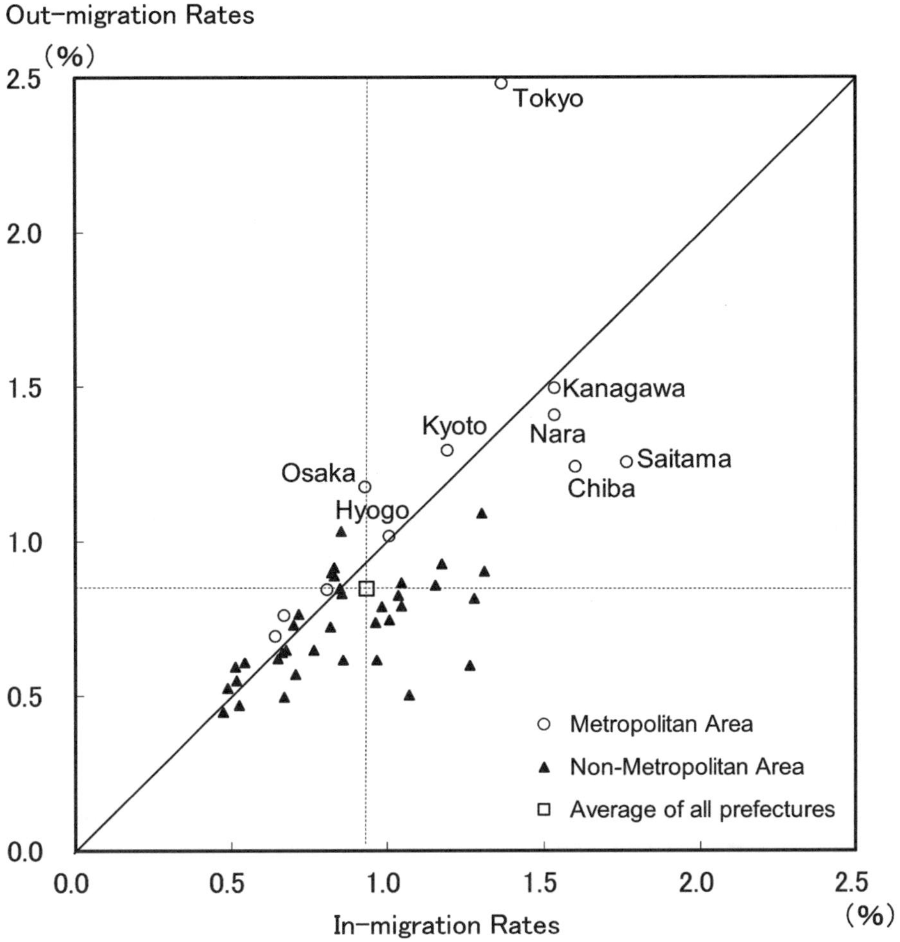

Fig. 7.1 Migration rates of elderly inter-prefectural migration by prefectures, Japan (2020). (Source: Population census)

and inactive areas in terms of both in- and out-migration. The dotted line in the figure indicates the average values of in-migration and out-migration rates for all prefectures. The prefectures in the upper right of the figure have a high gross migration rate (i.e., exceeding the national level for both out-migration and in-migration) and are areas with active elderly migration. The prefectures included in this part (Tokyo, Chiba, Kanagawa, Saitama, Kyoto, Hyogo, and Nara) belong to one of two metropolitan areas: the TMA and the Osaka Metropolitan Area. This part demonstrates that the inter-prefectural migration of older adults is active in metropolitan areas. In particular, the four prefectures in the TMA are plotted in the upper right corner and have high levels of both in- and out-migration compared to other prefectures.

Also, in the TMA, the out-migration rate is higher than the in-migration rate in Tokyo, indicating an excess of out-migration, while Kanagawa, Saitama, and Chiba prefectures have excess in-migration. The trends of the outflow of older adults from the central TMA and inflow to the suburbs of TMA are evident. In elderly migration, the migration rate in metropolitan areas is high, and the migration rate in non-metropolitan areas is low. Among the metropolitan areas, the city centers (Tokyo and Osaka Prefecture) demonstrate excess out-migration, and their suburbs have excess in-migration, as previously noted by Tahara and Iwadare [4] and Hirai [6], and a similar trend was confirmed in the 2020 data.

To investigate the characteristics of in- and out-migration in the TMA, we calculated the Migration Preference Index[4] on elderly inter-prefectural migration. Based on the analysis in the previous section, the migration of young-old migrants aged 65–74 years in 2020 (Fig. 7.2) and the migration of the old-old migrants aged 75 years and over in 2020 (Fig. 7.3) will be discussed separately. The total number of origin and destination pairs to exclude within-prefectural migrations is 2162 (47 × 46). Of these, 398 origin and destination pairs for young-old migration that involved a Migration Preference Index 100 and higher, the criterion for active migration, were observed. Of these, 103 originated from or were destined for one of the four prefectures of the TMA. For old-old migrations, there were 314 origin and destination pairs with a Migration Preference Index over 100, among which 80 pairs involved the TMA as an origin or destination. Both young-old and old-old migrants had a high Migration Preference Index for migration involving the TMA as a point of departure or arrival, suggesting their active migration patterns.

A characteristic of Japan is that migration within the same region or adjacent prefectures, such as in the Tohoku region and Kyushu region, is high and active and demonstrates a high Migration Preference Index. Still, the index is weak for migration to other regions. For example, there are no cases with a 100 or higher Migration Preference Index involving a move from a cold place such as the Tohoku region to the warmer regions of Kyushu. However, both young-old and old-old migrants TMA migrants have strong ties with neighboring areas (in the case of the TMA, Ibaraki, Tochigi, Gunma, Yamanashi, Nagano, and Shizuoka prefectures), as was observed with migrants of other regions. The Migration Preference Index exceeded 200 in most in- and out-migration cases between these regions. That is, approximately twice the volume of migrations occurred than the expected number of migrations.

The major difference with other regions is the relationship with other regions, most importantly, non-TMA areas. What characterizes migration involving

---

[4]The Migration Preference Index is the ratio of the expected number of migrations to the actual number of migrations, assuming that population migration occurs proportionately to the population size of the origin and destination [13]. A Migration Preference Index of 100 indicates an equal number of actual migrations as the expected number of migrations. A larger Migration Preference Index illustrates a higher number of actual migrations than the expected number of migrations, indicating active migration between the regions.

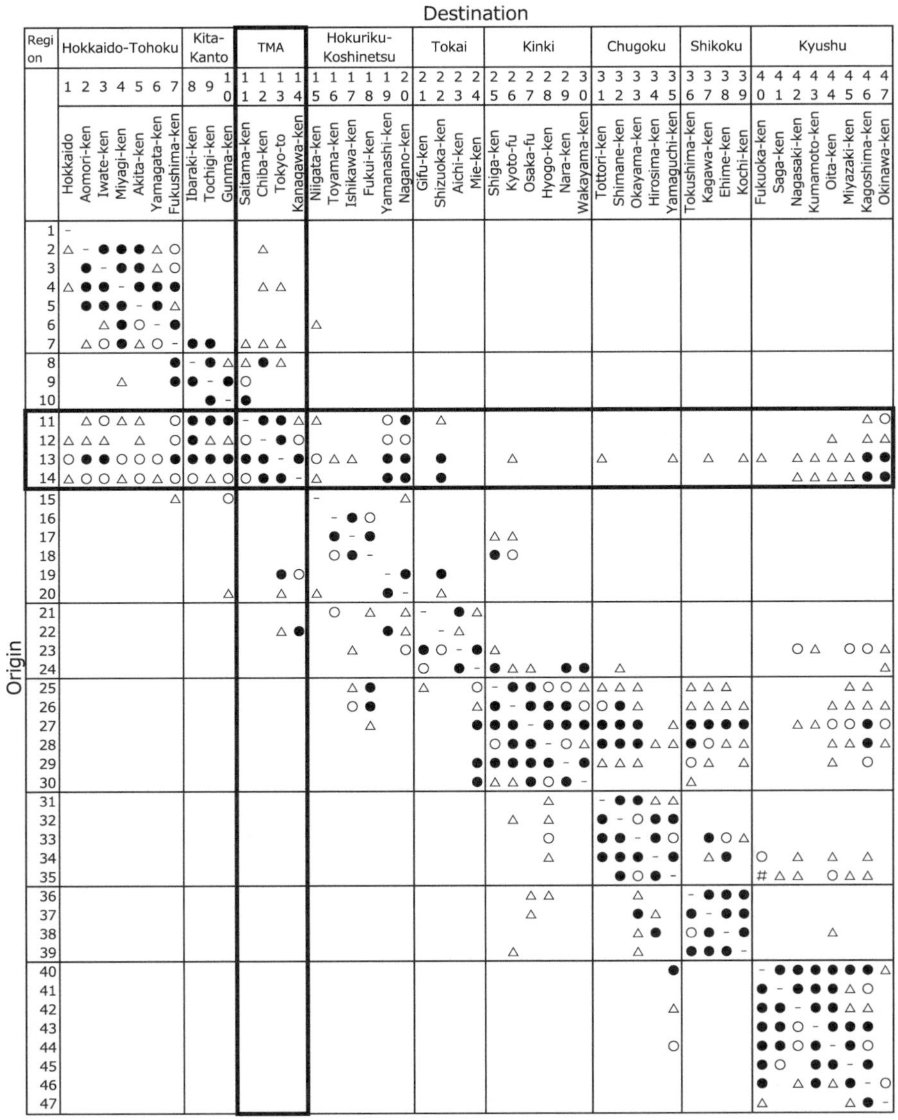

7 Elderly Migration in Tokyo Metropolitan Area, Japan

Fig. 7.2 Migration Preference Index of inter-prefectural migration by young-old (2020). Migration Preference Index △ 100 ~ 150, ○ 150 ~ 200, ● 200~. (Source: Population census). Note: *TMA* Tokyo Metropolitan Area

non-TMA regions is the gap between the young-old and the old-old. Among the young-old, there is active out-migration from the TMA to the Tohoku and Kyushu regions. In contrast, there is little in-migration from these regions, and out-migration to non-metropolitan areas dominates. Such an outflow of people from the TMA to Tohoku and Kyushu regions has also been noted in the 2000 Population Census

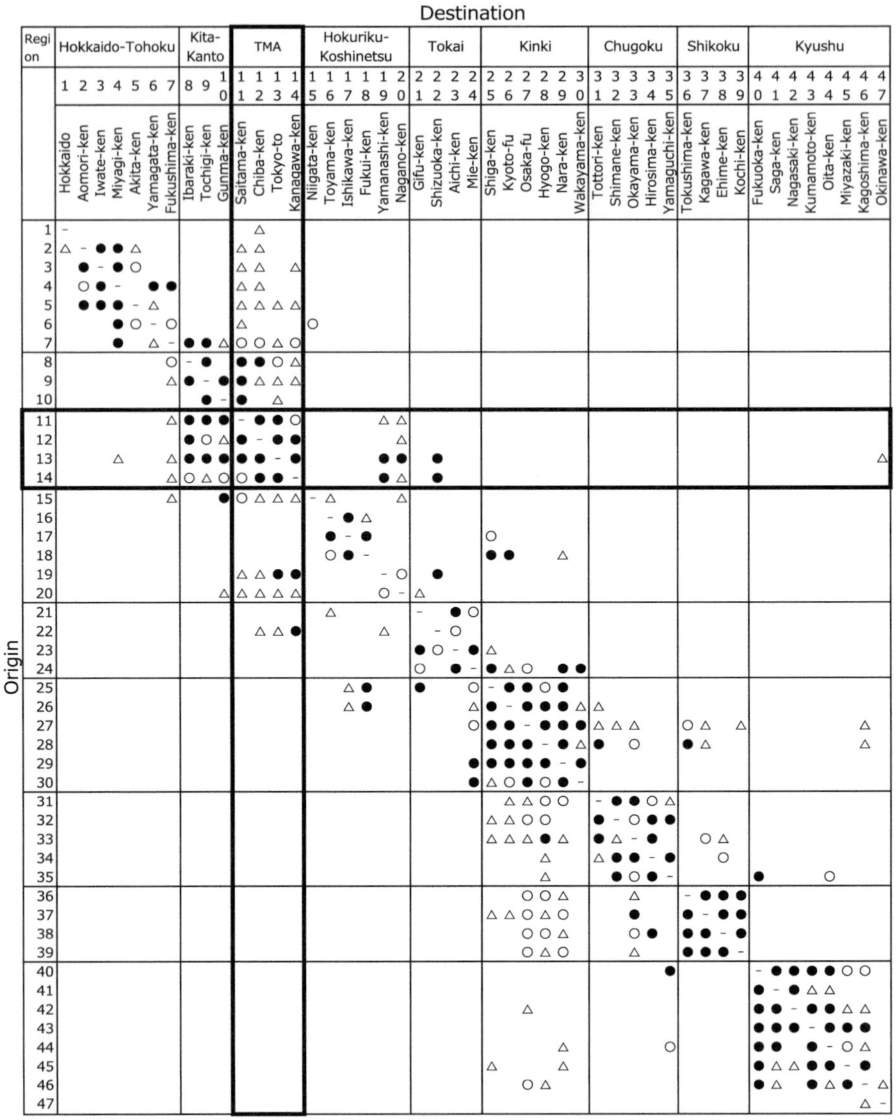

Fig. 7.3 Migration Preference Index of inter-prefectural migration by old-old (2020). Migration Preference Index △ 100 ~ 150, ○ 150 ~ 200, ● 200~. (Source: Population census). Note: *TMA* Tokyo Metropolitan Area

[6]. Still, this tendency has only gotten stronger in the 2020 data, where migration to the Kyushu region has become particularly notable. Among the old-old, there is little out-migration to non-metropolitan areas, whereas the Migration Preference Index is high for migration from the Tohoku and Kyushu regions, such as Nagasaki and Kagoshima prefectures, to the TMA.

Thus, migration to non-metropolitan areas, especially the Tohoku region, and the shift from excess outflow to excess inflow depending on age at the time of migration are the major characteristics of migration in the TMA.

7.5 Inter-Prefectural Migration in the TMA and Its Factors

So far, we have examined the characteristics of elderly migration in the TMA and have focused on inter-prefectural migration, which significantly changes the distribution of the Japanese older adult population. Based on the results, let us discuss the factors encouraging active inter-prefectural migration in the TMA.

First, young-old migrants flowed out of the TMA, notably to non-metropolitan areas such as the Tohoku and Kyushu regions. Few of these inter-prefectural migrants at this age worked, and the sex ratio was almost equal. This is likely an illustration of couples that relocate after retirement. The young-old migration from the TMA to non-metropolitan areas can reflect the reverse migration of young people who rushed to metropolitan areas from non-metropolitan areas during rapid economic growth. Therefore, it likely reflects migrants returning to their home prefectures. So far, migrations have been reported in remote Japanese islands such as Okinawa and Amami [7, 8] and communities that accept migrants have been discussed. Furthermore, the "retiree's return to farming," in which people move to rural areas and engage in farming after retirement, is also attracting attention [9]. The migration from the TMA to non-metropolitan areas pointed out in this chapter needs to be verified by a more detailed fact-finding survey. However, considering that the out-migration to non-metropolitan areas is even more prominent than according to the analysis of the 2000 data as the baby boomers that provided the large migration of rural populations into the metropolitan areas are retiring, it is expected that such migration will increase further.

Inter-prefectural migration of the old-old in the TMA was characterized by excess influx from non-metropolitan areas into the TMA. The inter-prefectural migration by the old-old consisted mostly of women, and most flowed into general households rather than institutional welfare households. Hirai [10], who analyzed the influx of older people to Tokorozawa City, Saitama Prefecture, describes a pattern in which widowed women move alone from various parts of Japan to live with their children who have settled in the TMA. The TMA has attracted many young people from other areas, but they are forced to leave their parents in their home regions and live separately when they settle in the TMA for employment or other purposes. Parents who used to live separately from their children moved into the households of their children living in metropolitan areas as they grew old and became single due to the death of their spouses.

Throughout its development and expansion, the TMA has attracted people from all over Japan, including non-metropolitan areas, and this trend continues today. This causes the out-migration of older adults who once flowed into the TMA back to non-metropolitan areas and, simultaneously, the in-migration from non-metropolitan areas to the children who settled in the TMA. The former represents migration from the TMA to rural areas that occurs relatively early in old age, and the latter consists of migration to the TMA that occurs later in life. The concurrence of these two types of migration, involving different age groups and migration directions, has made inter-prefectural migration in the metropolitan area more active, giving rise to the unique characteristics that distinguish the TMA from other regions.

7.6 Conclusions

This chapter investigated the characteristics of elderly migration in the TMA. The TMA was the earliest location in Japan to observe elderly migration and continues to be the most active area in terms of rate and absolute numbers.

Elderly migration in the TMA is characterized by a comparatively high rate of inter-prefectural migration than in other areas and greatly impacts the distribution of the elderly population in Japan as a whole. Inter-prefectural migration in the TMA is high for both in- and out-migration, but trends change according to migrant age. Compared to the young-old migrants who migrated more often from the TMA to non-metropolitan areas such as the Tohoku or Kyushu regions, the reverse movement of non-metropolitan areas to the TMA was notable among old-old migrants. This is thought to be related to the characteristics of the TMA that have traditionally drawn in a large population and are the very characteristics of elderly migration in the TMA.

Based on these facts, it will be necessary to pay attention to the following in the future as the TMA population ages. The aging of the baby boomers will rapidly age the TMA population, but the TMA also has an active migration of older adults than other regions of Japan. Under such circumstances, it is an urgent task to create a system to accommodate incoming elderly migrants. Moving to a new residence and adapting to a new living environment can have negative physical and mental health impacts on older adults [11], so it is necessary to provide sufficient care for migrants. This is particularly true regarding accommodating the inter-prefectural influx of older adults, and this effort may require cooperation with medical institutions in their former areas of residence. Furthermore, the influx from non-metropolitan areas is also expected to further increase the demand for welfare services in the TMA. Appropriate services must be provided after collecting accurate data on the realities of elderly migration.

Acknowledgments This chapter is based on the following original paper published in Japanese, except that the data has been updated: Hirai M (2014) Kourei jinko ido (Elderly Migration). In: Inoue T. and Watanabe M. (eds) Syutoken no koureika (Population Aging in Tokyo Metropolitan Area), Hara Shobo, Tokyo, 53-71. I am grateful to Hara Shobo for permitting me to translate and republish the paper.

References

1. Rogers A, Woodward J. The sources of regional elderly population growth: migration and aging-in-place. Prof Geogr. 1988;40:450–9.
2. Otomo A. Mobility of elderly population in Japanese metropolitan areas. J Popul Stud. 1981;4:23–8.
3. Uchino S. Elderly migration in Japan—newly emerging trend and its analysis. J Popul Probl. 1987;184:19–38.
4. Tahara Y, Iwadare M. Where the elderly move to: a review and a study of elderly migration flows in Japan. Komaba Stud Hum Geogr. 1999;13:1–53.
5. Otomo A. Forms of migration in old age. Municipal Probl. 1999;90(12):17–28.
6. Hirai M. Regional characteristics of elderly inter-prefectural migration. In: Ishikawa Y, editor. Population decline and regional imbalance geographic perspectives. Kyoto: Kyoto University Press; 2007. p. 129–47.
7. Tahara Y, Nagata J, Arai Y. Investigation of the process and effects of elderly return migration—Case of Settlement N. Jpn J Gerontol. 2000;22:436–48.
8. Suyama S, Jung MA. Residential migration patterns of inhabitants in Naze, Amami-Oshima Island; cases of households coming from Odana Hamlet. Regional views (Komazawa University). Inst Appl Geogr. 2004;17:81–95.
9. Takahashi I. Organizational support of retiree returning farmers and its development: case of Suou oshima chou in Yamaguchi prefecture. Month Bull Agric Stat Res. 2005;2005(5):41–7.
10. Hirai M. Characteristics of in-migration of the elderly in a suburb of Tokyo: a case study of Tokorozawa City, Saitama Prefecture. Geogr Rev Japan. 1999;72A:289–309.
11. Ando T, Koyano W, Yatomi N, Watanabe S, Kumagai O. Migration of community-dwelling older adults and adaptation after migration. Jpn J Gerontol. 1995;16:172–8.
12. Hirai M. Previous household composition and residence of elderly residents in a nursing home in Okutama, Tokyo. J Popul Stud. 2000;27:15–22.
13. Bachi R. Statistical analysis of geographic series. Bull Inst Int Stat. 1958;36(2):229–40.

Chapter 8
Simplified Projection of the Insurance Premiums in the Greater Tokyo Area, 2020–2060

Nozomu Inoue

Abstract This study projected the future insurance premiums from 2020 to 2060 in the Greater Tokyo Area, given the expected increasing demand for nursing care. Estimating the change in the mean insurance premium rates by prefecture showed that in almost all prefectures, insurance premiums will reach their peak in Phase XIV (2040); however, in Gunma Prefecture, they will reach their peak in Phase VII (2020), and in Tokyo Prefecture, they will reach their peak in 2050. Next, verifying the index of increase in premiums by insured person across the entire Greater Tokyo Area from 2018 to 2060, calculated with the Phase V premium rates as 100, showed that over the entire duration, the kurtosis of the distribution tended to decrease gradually, while its variance tended to increase. It also showed an increase in the number of regions where the index became negative. Meanwhile, after 2020, there were also some regions where the indices became unusually high. In addition, verifying the distribution of the index of increase in premiums by insured person by prefecture showed that the variance of each prefecture's distribution increased, just like that of the distribution across the Greater Tokyo Area.

Keywords Projection · Insurance premiums · Greater Tokyo Area

8.1 Introduction

Besides a reduction in the working population, Japan's aging society and declining birthrate suggest a host of other problems. Among these is the demand for nursing care. Japan's longevity rate is one of the world's highest. According to the Ministry of Health, Labor, and Welfare [1], the mean life expectancy was over 80 years for both sexes in 2016 (80.98 years for men and 87.14 years for women). However, increased longevity is a primary factor in the aging of society; that is, prolonging the mean life expectancy can play a part in accelerating the proportion of people over 65 in the population (the population aging rate). Figure 8.1 shows the distribution of the population aging rate of various small areas in the Greater Tokyo Area, created using data from Inoue [2, 3]. The figure makes it clear that the population aging rate is increasing yearly, and in 2060, it will also be high in Tokyo's 23 wards.

Based on this result, the number of elderly people in the Greater Tokyo Area is expected to increase, which suggests that the number of people in need of nursing care should also increase. Inoue [4] shows how many people will be in need of nursing care in the Greater Tokyo Area in the future and that the provision of locations for current nursing-care facilities is insufficient, particularly in the city center. Increased demand for nursing care will affect not just locations for nursing-care facilities but insurance premiums as well. Insurance premiums are determined based on the Insured Long-Term Care Service Plans, formulated every three years. They cover approximately 90% of the expenses of nursing-care services for people who qualify as needing nursing care, of which approximately 50% each is covered by public funds and by premiums collected from people with nursing-care insurance. In addition, of the approximately 50% covered by this premium, about 22% is the total of the premiums paid by primary insured people. Furthermore, dividing this total of premiums by the number of primary insured people gives the base premium rate. Therefore, an increase in the demand for nursing care may play a part in increasing insurance premiums.

Thus, given an increase in the elderly population and consequent increase in people in need of nursing care, it is of utmost importance to examine how much insurance premiums can be expected to increase. This is because if there are areas that see an extreme increase in insurance premiums, this may result in primary insured people who cannot pay the high premiums. If this happens, it will cause a negative chain reaction, as other resources or other insured people's premiums will have to be used to compensate, which may cause a further increase in insurance premiums.

To that end, the objective of this study is to project the future insurance premiums from 2020 to 2060 in the Greater Tokyo Area, where the elderly population is expected to increase, to identify (1) future trends in insurance premiums and (2) which regions are expected to see a particularly sharp increase in insurance premiums. First, Sect. 8.2 will describe the data used in this study and the model used to project future insurance premiums. Section 8.3 will explain the results of the actual

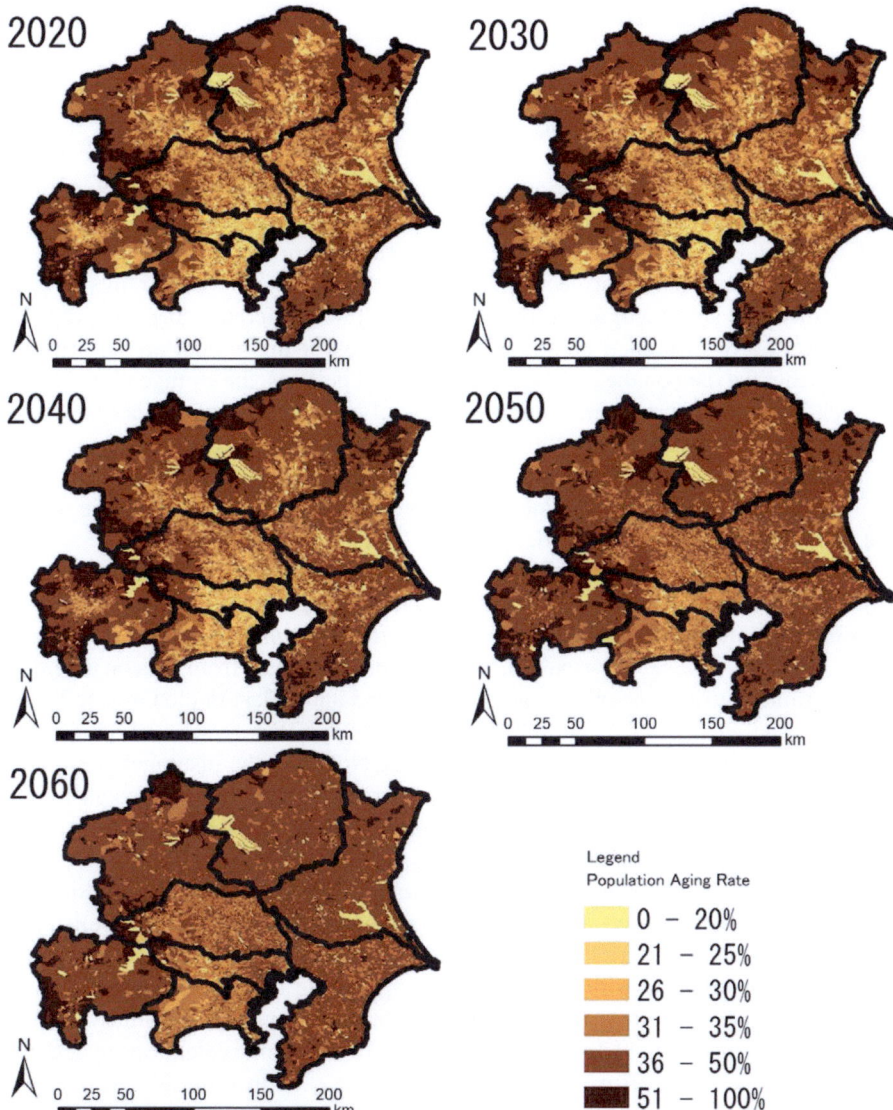

Fig. 8.1 Distribution of the population aging rate of various small area in the Greater Tokyo Area, 2020–2060

estimation of future insurance premiums. Finally, Section 8.4 will summarize the discussion up to that point and mention future research tasks.

Note that "Greater Tokyo Area" in this study refers to the Capital Region (*shutoken*) as defined by the Capital Region Arrangement Act, which is Tokyo and the seven prefectures of Kanagawa, Chiba, Saitama, Yamanashi, Gunma, Tochigi, and Ibaraki (rather than only Kanagawa, Chiba, and Saitama).

8.2 Data and Method of Analysis

As stated in the previous section, this study projects future insurance premiums from 2020 to 2060 in the Greater Tokyo Area, given the increasing demand for nursing care expected in the future, to identify (1) future trends in insurance premiums and (2) which regions are expected to see a particularly sharp increase in insurance premiums.

Insurance premiums are normally calculated using the Ministry of Health, Labor, and Welfare's "Worksheet for Insured Long-Term Care Service Plans," however much data is required to project insurance premiums. Due to constraints in data availability, this study will use a simplified model, with reference to Yamamoto [5], to project future insurance premiums. The features of this model are as follows: suppose that the data on the number of first insured persons, the storage rate of long-term care insurance premiums, the proportion of insured persons by income stage, and other data that exclude population data are constant throughout the future, the number of data required for calculation was reduced. Therefore, the results of the trial calculation were also examined on the premise that the present and the rest of the population remain unchanged.

The data used in the analysis are the *number of primary insured people*, the *adjusted number of primary insured people*, *base-year premiums*, the *rate of receipt of insurance premiums*, and *charges to be borne by primary insured people* per capita. First, the *number of primary insured people* was taken from data on the number of primary insured people (population aged 65 and older) in the Greater Tokyo Area by small area from 2020 to 2060 in Inoue's "The Web System of Small Area Population Projection for the Whole Japan" [2]. Note that the insurance premiums are defined for each insured person (i.e., first, second), so data segregated by the insured person were used for future projects after totaling the number of primary insured people by small area.

Next, the *adjusted number of primary insured people* was obtained by using the number of primary insured people (by income level) in 2015 from the Ministry of Health, Labor, and Welfare [6] data (which is segregated by insured person) to calculate an adjustment factor, which was then applied to the future population as a constant. To give a concrete example, Fig. 8.2 shows the number of primary insured people in Chiyoda Ward in Tokyo Prefecture, divided into nine categories by income level. Next, the *share of primary insured people* at each income level was calculated. These shares would be used to project the future *adjusted number of primary insured people*. The *number of primary insured people* at each income level was then multiplied by the premiums defined by the Ministry of Health, Labor, and Welfare for each income level. Finally, the result from multiplying the *number of primary insured people* at each income level by the corresponding premium was summed for all rates to calculate the *adjusted number of primary insured people* in Chiyoda in Tokyo Prefecture, 2015. Similarly, the *number of primary insured people* in 2020, which is a projected value, was multiplied by the *share of primary insured people* at each income level that was just calculated. Those values were again multiplied by the rates for each income level, and the total sum of those values is the *adjusted number of primary insured people* in 2020, as shown in Fig. 8.3.

8 Simplified Projection of the Insurance Premiums in the Greater Tokyo Area... 109

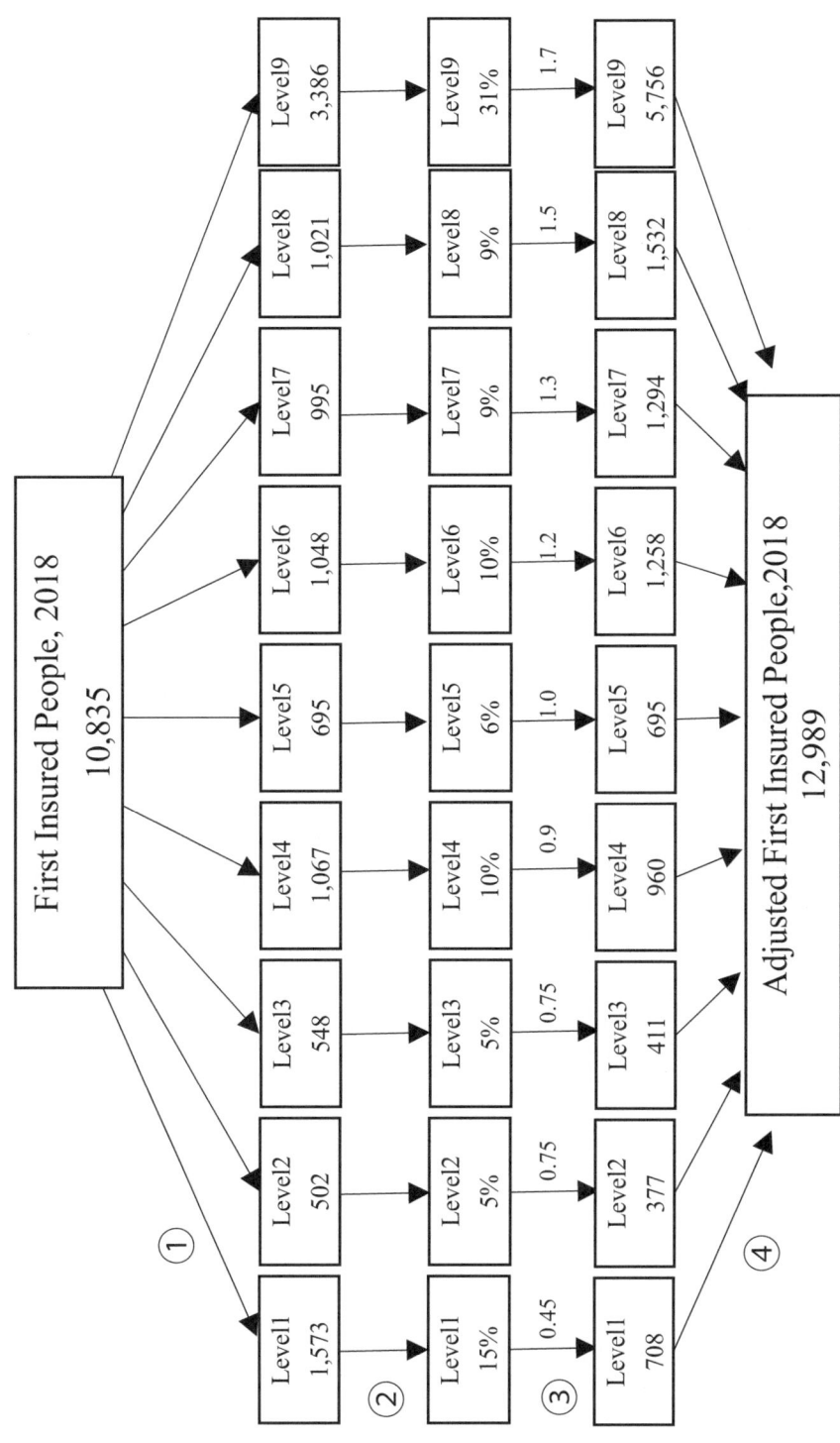

Fig. 8.2 How to adjust the number of insured people of Chiyoda in Tokyo, 2018

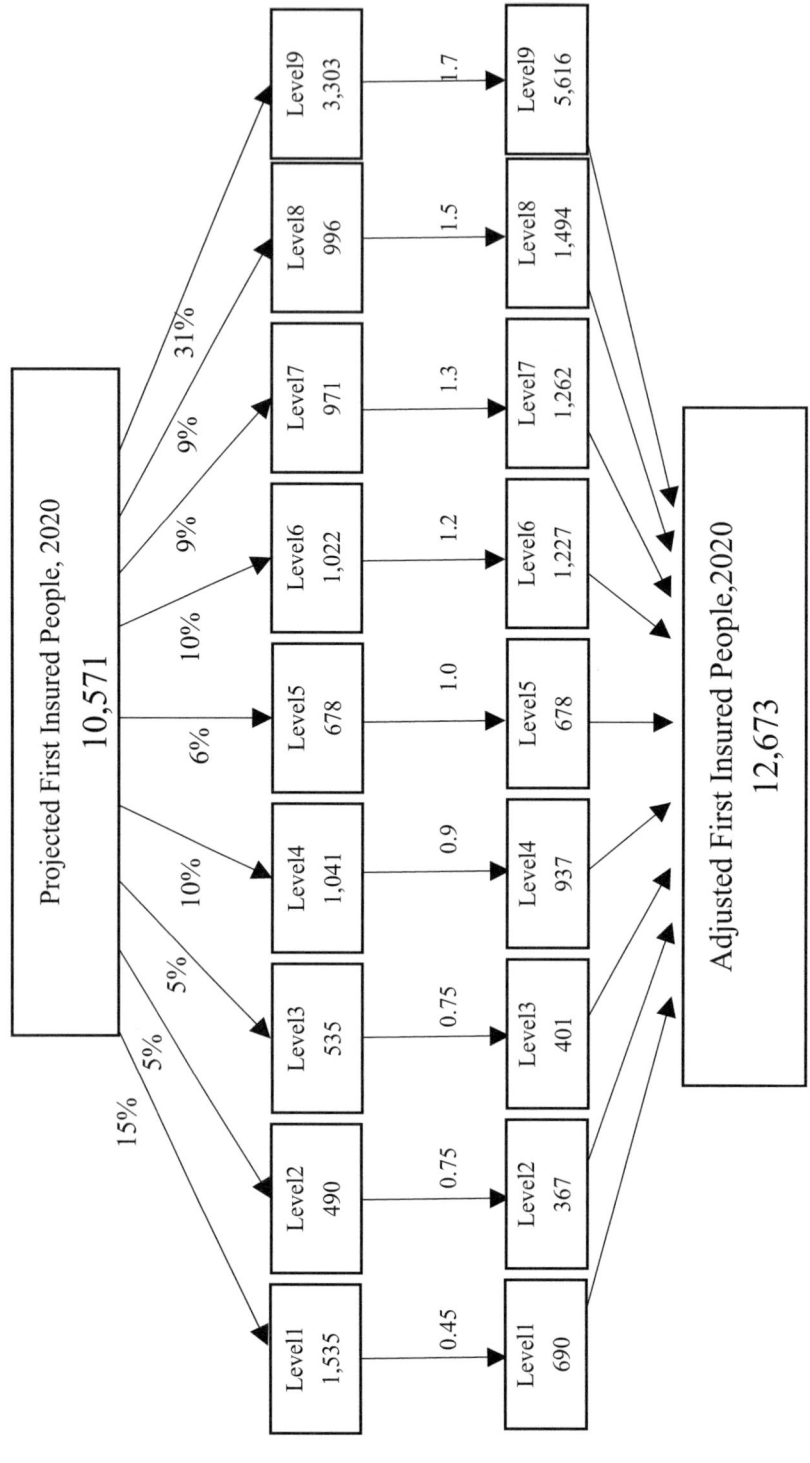

Fig. 8.3 How to adjust the number of insured people of Chiyoda in Tokyo, 2020

Moving on, a summary of the data from the Ministry of Health, Labor, and Welfare [7, 8] on the base premium rate for each insured person was used as the data for the *base-year premiums*. The base premium rate for each insured person is calculated based on the Phase VII (newest) Insured Long-Term Care Service Plan. The *rate of receipt of insurance premiums* was calculated by dividing the insurance premiums received by the insurance premiums scheduled, with reference to Matsuoka and Nakazawa [9]. Finally, *charges to be borne by primary insured people* per capita were treated as an unknown variable and calculated from the equations of the following simplified model.

The model for the insurance premiums in Yamamoto [5] is as in Eq. (8.1).

$$\text{LTCpremium}_i = \frac{\text{LTCcost}_i \times \text{LTCpop}_i}{\text{Adjusted LTCpop}_i \times r_i \times 12} \qquad (8.1)$$

Then, we set i as the number of insurers, LTCpremium as insurance premiums, LTCcost as charges to be borne by primary insured people per capita, LTCpop as the number of primary insured people, *adjusted* LTCpop$_i$ as adjusted number of primary insured people, r as the pay rate of long-term care insurance premium. Here, LTCcost was an unknown variable, and it was estimated with Eq. (8.2).

$$\text{LTCcost}_i = \frac{\text{LTCpremium}_i \times \text{Adjusted LTCpop}_i \times r_i \times 12}{\text{LTCpop}_i} \qquad (8.2)$$

Future insurance premiums were projected using the estimated variable in Eq. (8.3). Then, we set t as the year of projection, and other variables are the same as above. Furthermore, the data of the phase VII Insured Long-Term Care Service Plan shall be used assuming that the burden amount of the primary insured person per capita and the pay rate of nursing care insurance premiums are constant in estimating.

$$\text{LTCpremium}_i(t) = \frac{\text{LTCcost}_i(2018) \times \text{LTCpop}_i(t)}{\text{Adjusted LTCpop}_i(t) \times r_i(2018) \times 12} \qquad (8.3)$$

8.3 Simplified Projection of the Insurance Premiums

Section 8.2 describes the data necessary to make simplified projects of the future insurance premiums in the Greater Tokyo Area and the data analysis method. This section will verify those projects to identify (1) future trends in insurance premiums and (2) which regions are expected to see a particularly sharp increase in insurance premiums.

First, Fig. 8.4 shows the mean insurance premium rates by prefecture up to the year 2060. While insurance premiums in nearly all prefectures will be at their peak

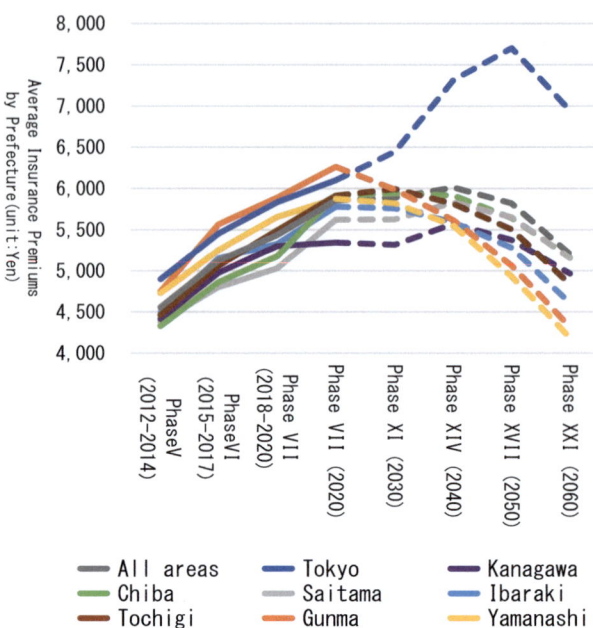

Fig. 8.4 Trends in average insurance premiums by prefecture, 2012–2060

Table 8.1 Trends in the over-65 population of the Greater Tokyo, 2010–2060

	2010	2020	2030	2040	2050	2060
				(Unit: 1000 people)		
All areas	9,114	11,457	11,867	13,201	13,507	12,247
Tokyo	2,684	3,195	3,370	3,986	4,349	4,045
Kanagawa	1,829	2,324	2,453	2,799	2,842	2,605
Chiba	1,338	1,788	1,856	2,038	2,060	1,846
Saitama	1,469	1,930	1,963	2,141	2,127	1,913
Ibaraki	667	843	847	852	819	713
Tochigi	442	557	566	570	549	474
Gunma	473	575	566	571	537	459
Yamanashi	213	245	245	244	223	190

in 2040 (Phase XIV), Only Gunma and Tokyo Prefectures exhibit different changes. The mean rate in Gunma Prefecture will be at its peak in 2020 (Phase VII), and the mean rate in Tokyo Prefectures will be at its peak in 2050. Of particular note is that the peak Tokyo Prefecture rate exceeds ¥7500 according to the trial calculation. This difference in peaks can be attributed to the difference in the rate at which the population ages from prefecture to prefecture. In other words, Gunma Prefecture is expected to age sooner than other prefectures, while Tokyo Prefecture is expected to age later than other prefectures. Table 8.1 shows the trends in the over-65 populations of the Greater Tokyo Area. As this table shows, there will be almost no change in Gunma Prefecture from 2020 to 2040, and this phase is at its peak. The table also

8 Simplified Projection of the Insurance Premiums in the Greater Tokyo Area... 113

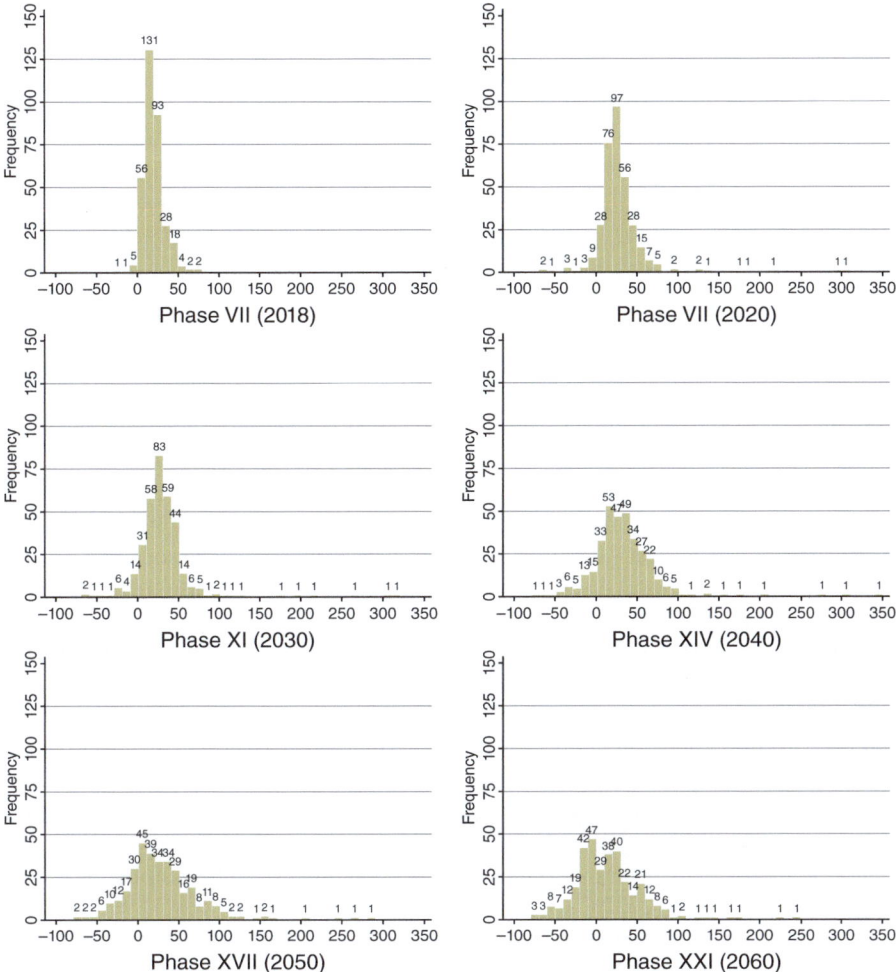

Fig. 8.5 Distribution of the index of increases in premiums by insured person across the entire Greater Tokyo Area, 2018–2060. Note: Phase 5 nursing care insurance premium amount was set to 100

confirms that Tokyo Prefecture will reach its peak in 2050. As the population of each prefecture will age at a different rate, insurance premiums will not peak uniformly but will exhibit different changes from one prefecture to the next.

Next is a more detailed projection. Figure 8.5 shows the distribution of the index of increases in premiums by insured persons across the entire Greater Tokyo Area from 2018 to 2060. This index of increase was calculated with the Phase V premium rates as 100. Note that the Phase VII premium rates announced by the Ministry of Health, Labor, and Welfare are written alongside the rates calculated using the projected population data in this study. Over the entire duration, the kurtosis of the distribution tends to decrease gradually while its variance tends to increase. The mean values are 19.9 in Phase VII (2018), 29.1 in Phase VII (2020), 29.8 in Phase XI (2030), 32.1 in Phase XIV (2040), 27.7 in Phase XVII (2050), and 14.2 in Phase XXI (2060), which matches the results in Fig. 8.4, where the peak occurs in Phase XIV. It can also be observed that there is an increase in the number of regions where the index becomes negative—that is, the rates of their premiums fall below those of Phase V. Meanwhile, after 2020, there are also some regions where the indices become unusually high. It is supposed that this occurs in regions where the number of insured people was extremely small, to begin with. The reason for this is that in such regions, a decrease by even one person could result in a large rate of decrease, so insurance premiums projected from the number of insured people are likewise influenced similarly.

After that, we confirmed the distribution of the index of increase in premiums by insured person by prefecture. Figures 8.6, 8.7, 8.8, 8.9 and 8.10 show the distribution of the index of increase in premiums by insured person by prefecture from 2020 to 2060. Due to the small sample size, the distributions have an irregular form, unlike that of the entire Greater Tokyo Area. However, the variance in each prefecture's distribution will still increase in the same way. It can also be seen that Tokyo, Saitama, and Chiba Prefectures have the highest number of regions where the index becomes unusually high.

Next, we identified the regions with an unusually high index of increase in premiums by insured person. Ordinarily, a Smirnov-Grubbs test should be used to calculate statistically significant anomalies. However, for this study, anomalies were defined as cases where the index exceeded 150; that is, the rates had increased to 1.5 times the Phase V premium rates. Then, this definition of anomaly was used to confirm the anomalous regions by fiscal year. Note that in no region did the index ever go below −100. Table 8.2 presents an overview of the number of regions where the index of increase in insurance premiums was anomalous. The number was five regions in Phase VII (2020); six regions in Phase XI (2030); six regions in Phase XIV (2040); seven regions, the most, in Phase XVII (2050); and four regions, the fewest, in Phase XXI (2060). In each of these phases, the anomalous regions were always in Saitama, Chiba, or Tokyo Prefecture. The only anomalies in Tokyo Prefecture were the villages of Aogashima and Mikurashima; there were none on the mainland. The highest index in all phases was 346 in Aogashima during Phase XIV (2040), and the regions that were anomalous for the entire duration were Tōnoshō, Chiba Prefecture; Higashi-chichibu, Saitama Prefecture; and Kamikawa, Saitama Prefecture. The highest number of insured people was 65,663 in Urayasu, Chiba Prefecture, in Phase XVII (2050), and the lowest was 43 in Aogashima, Tokyo Prefecture, in Phase XXI (2060), when there were few insured people in every region.

Finally, the last step was to see if there were any geographical characteristics in the insurance premiums for each insured person. Figure 8.11 shows the distribution of the insurance premiums in the Greater Tokyo Area by insured person. Whereas Table 8.2 highlights the regions where the index of increase is unusually high, Fig. 8.11 is a choropleth that represents regions with high premiums as actual

8 Simplified Projection of the Insurance Premiums in the Greater Tokyo Area… 115

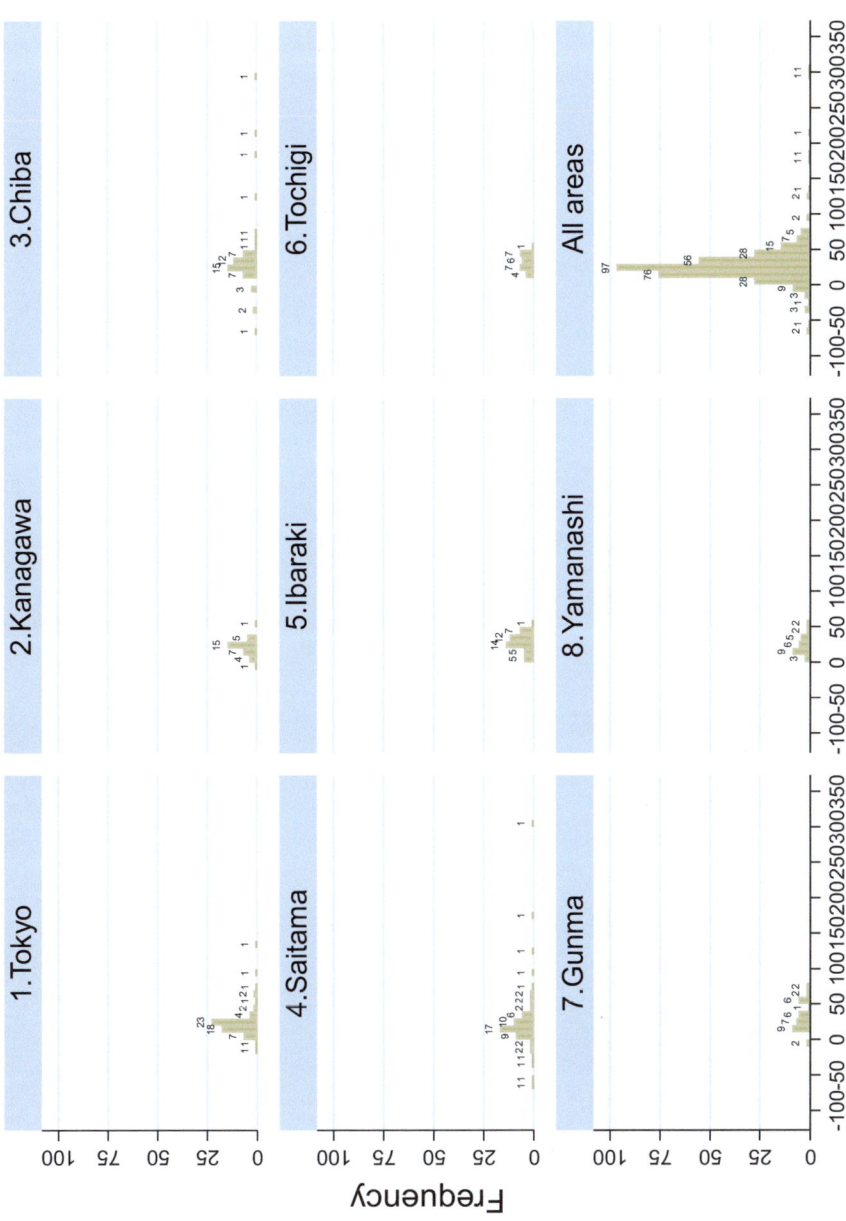

Fig. 8.6 Distribution of the index of increase in premiums by insured person by prefecture, Phase VII (2020)

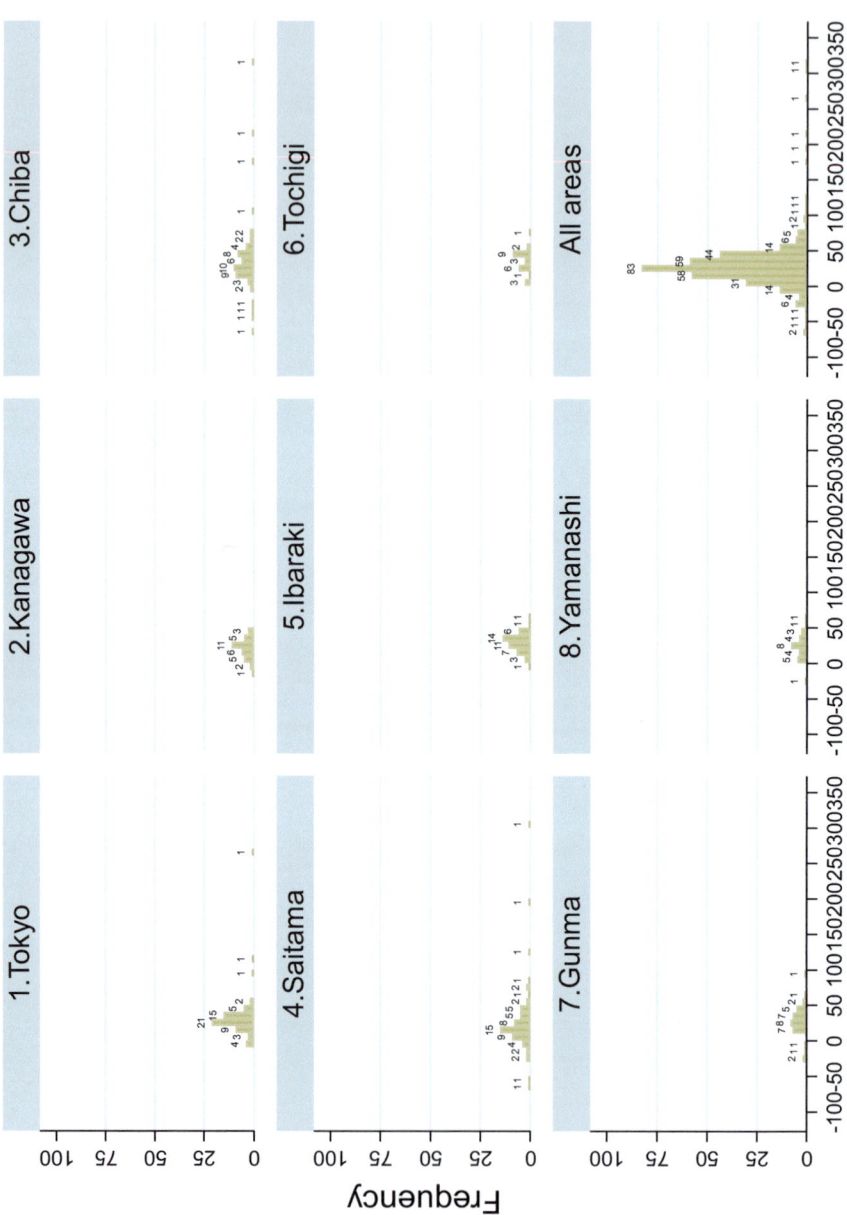

Fig. 8.7 Distribution of the index of increase in premiums by insured person by prefecture, Phase XI (2030)

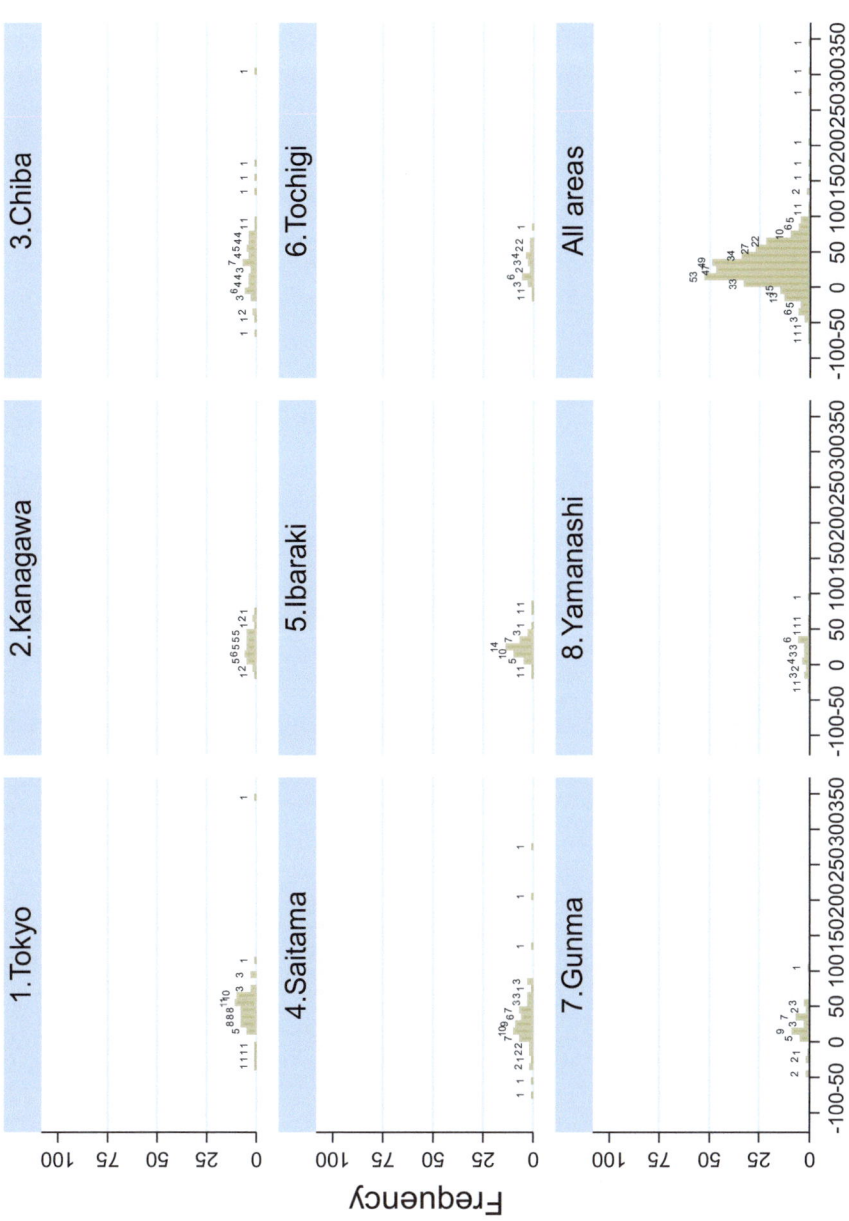

Fig. 8.8 Distribution of the index of increase in premiums by insured person by prefecture, Phase XIV (2040)

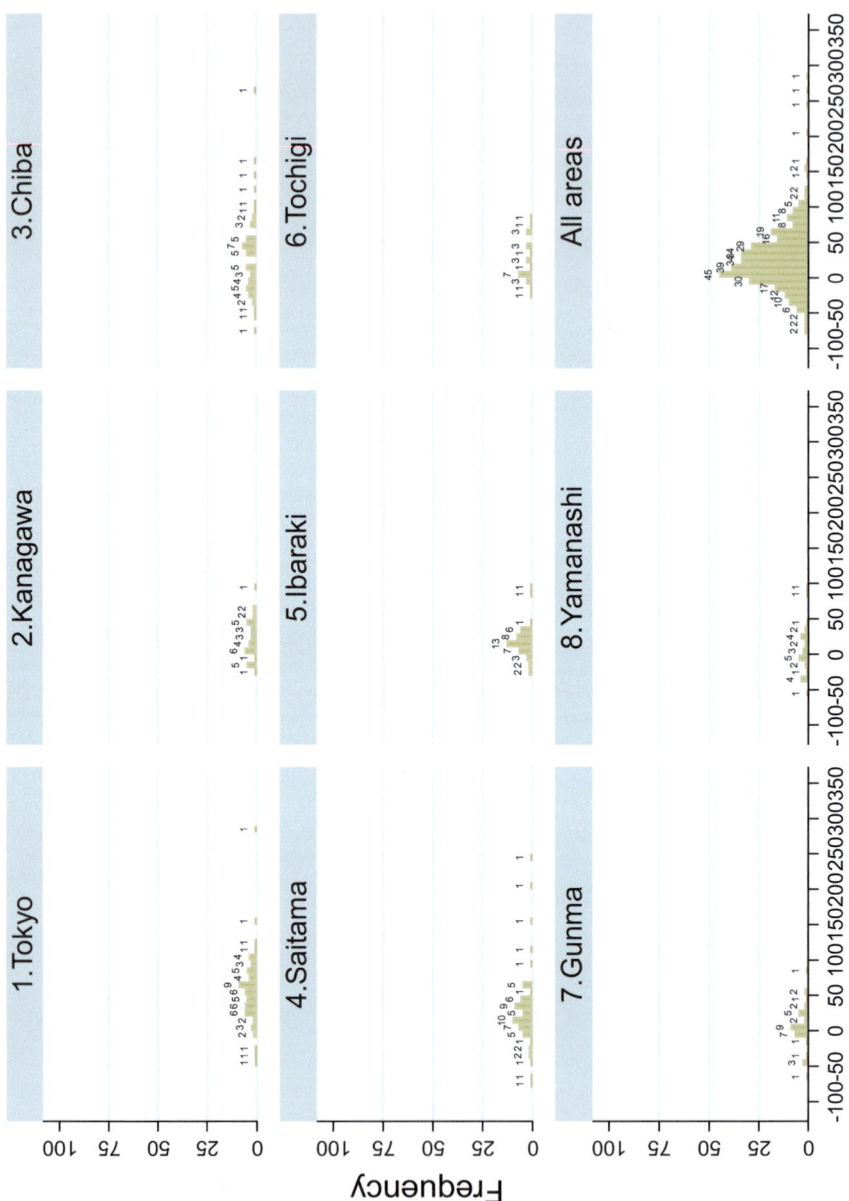

Fig. 8.9 Distribution of the index of increase in premiums by insured person by prefecture, Phase XVII (2050)

8 Simplified Projection of the Insurance Premiums in the Greater Tokyo Area...

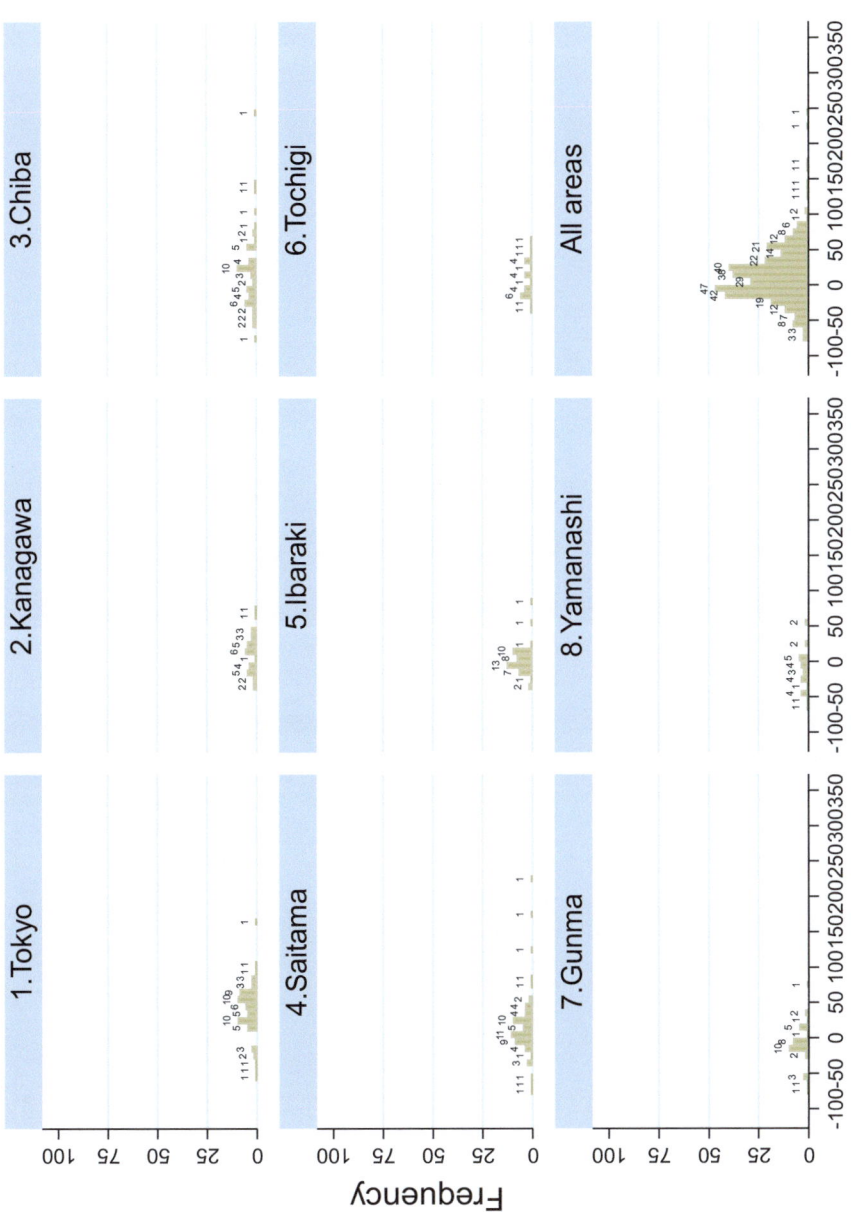

Fig. 8.10 Distribution of the index of increase in premiums by insured person by prefecture, Phase XXI (2060)

Table 8.2 Overview of the number of regions where the index of increase in the insurance premiums was anomalous

Phase VII (2020)

Pref.	City	Index of increase	The number of care insured	Insurance premiums
Saitama	Higashi-chichibu	302	3478	22,003
Chiba	Tōnoshō	294	15,682	15,941
Chiba	Kōzaki	214	5,377	12,574
Chiba	Isumi	182	28,163	11,507
Saitama	Kamikawa	179	8,111	12,675

Phase XI (2030)

Pref.	City	Index of increase	The number of care insured	Insurance premiums
Chiba	Tōnoshō	319	16,701	16,977
Saitama	Higashi-chichibu	304	3,489	22,073
Tokyo	Aogashima	266	60	20,154
Chiba	Kōzaki	211	5,318	12,436
Saitama	Kamikawa	194	8,573	13,397
Chiba	Isumi	178	27,761	11,343

Phase XIV (2040)

Pref.	City	Index of increase	The number of care insured	Insurance premiums
Tokyo	Aogashima	346	73	24,521
Chiba	Tōnoshō	304	16,089	16,355
Saitama	Higashi-chichibu	278	3,266	20,662
Saitama	Kamikawa	207	8,926	13,948
Chiba	Kōzaki	179	4,768	11,150
Chiba	Isumi	155	25,462	10,404

Phase XVII (2050)

Pref.	City	Index of increase	The number of care insured	Insurance premiums
Tokyo	Aogashima	285	63	21,162
Chiba	Tōnoshō	266	14,587	14,828
Saitama	Higashi-chichibu	241	2,950	18,663
Saitama	Kamikawa	202	8,797	13,747
Chiba	Urayasu	169	65,663	11,041
Tokyo	Mikurashima	157	123	10,468
Saitama	Toda	151	41,956	11,119

Phase XXI (2060)

Pref.	City	Index of increase	The number of care insured	Insurance premiums
Chiba	Tōnoshō	245	13,744	13,971
Saitama	Higashi-chichibu	223	2,795	17,682
Saitama	Kamikawa	178	8,103	12,662
Tokyo	Aogashima	163	43	14,444

Note: An abnormal value was set when the increase index of the insurance premium exceeded 150

8 Simplified Projection of the Insurance Premiums in the Greater Tokyo Area...   121

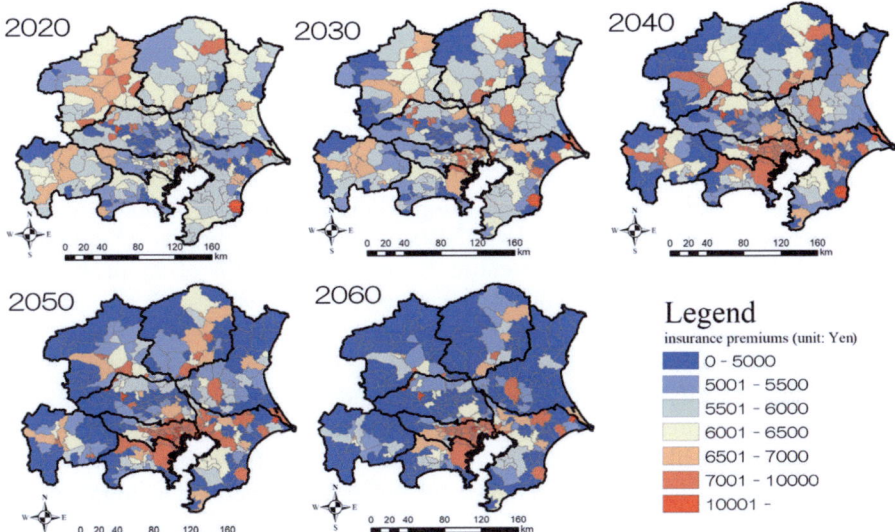

Fig. 8.11 The Distribution of the insurance premiums in the Greater Tokyo Area by insured person

numbers. Normally, actual population numbers would be represented with graduated symbols rather than a choropleth map, but a choropleth map is used here as the numbers are the per capita premiums. First, by Phase VII (2020), there is already a scattering of regions with high premiums, such as northwestern Gunma Prefecture; Odawara, Tochigi Prefecture; Isumi, Chiba Prefecture; and Higashi-chichibu, Saitama Prefecture. This trend is the same in 2030, except in the 23 wards of Tokyo, where there are significant changes. The premiums in nearly all of the 23 wards of Tokyo increase, and this trend continues until 2060. By 2040, premiums can also be seen to increase in regions such as Yokohama, Kanagawa Prefecture; Kawasaki, Kanagawa Prefecture; Chiba, Chiba Prefecture; and Funabashi, Chiba Prefecture—a remarkable trend in the so-called city center environs. Meanwhile, by 2060, the insurance premiums have decreased significantly in almost all regions of Gunma and Yamanashi Prefectures, indicating an increasing bipolarization of insurance premiums.

8.4   Conclusion

This study projected the future insurance premiums from 2020 to 2060 in the Greater Tokyo Area, given the increasing demand for nursing care expected in the future. This was done to identify (1) future trends in insurance premiums and (2) which regions are expected to see a particularly sharp increase in insurance premiums. Insurance premiums are normally calculated using the Ministry of Health, Labor, and Welfare's "Worksheet for Insured Long-Term Care Service Plans," but due to constraints in the availability of data, this study used a simplified model, with reference to Yamamoto [5], to project future insurance premiums.

Estimating the change in the mean insurance premium rates by prefecture showed that in almost all prefectures, insurance premiums will reach their peak in Phase XIV (2040); however, in Gunma Prefecture, they will reach their peak in Phase VII (2020), and in Tokyo Prefecture, they will reach their peak in 2050.

Next, verifying the index of increase in premiums by insured person across the entire Greater Tokyo Area from 2018 to 2060, calculated with the Phase V premium rates as 100, showed that over the entire duration, the kurtosis of the distribution tended to decrease gradually, while its variance tended to increase. It also showed an increase in the number of regions where the index became negative. Meanwhile, after 2020, there were also some regions where the indices became unusually high.

In addition, verifying the distribution of the index of increase in premiums by insured person by prefecture showed that the variance of each prefecture's distribution increased, just like that of the distribution across the entire Greater Tokyo Area.

Moving on, the regions with an unusually high index of increase in premiums by insured persons were identified.

The number of regions where the index was anomalous was five regions in Phase VII (2020); six regions in Phase XI (2030); six regions in Phase XIV (2040); seven regions, the most, in Phase XVII (2050); and four regions, the fewest, in Phase XXI (2060). The highest index in all phases was 346 in Aogashima during Phase XIV (2040), and the regions that were anomalous for the entire duration were Tōnoshō, Chiba Prefecture; Higashi-chichibu, Saitama Prefecture; and Kanagawa, Saitama Prefecture.

Here, the insurance premiums for each insured person were tested for any geographical characteristics. First, by Phase VII (2020), there was already a scattering of regions with high premiums, such as northwestern Gunma Prefecture; Odawara, Tochigi Prefecture; Isumi, Chiba Prefecture; and Higashi-chichibu, Saitama Prefecture. This trend was the same in 2030, except in the 23 wards of Tokyo, where there were significant changes. The premiums in nearly all of the 23 wards of Tokyo increased, and this trend continued until 2060. By 2040, premiums could also be seen to increase in regions such as Yokohama, Kanagawa Prefecture; Kawasaki, Kanagawa Prefecture; and Chiba, Chiba Prefecture—a remarkable trend in the so-called city center environs. Meanwhile, by 2060, insurance premiums had decreased significantly in almost all regions of Gunma and Yamanashi Prefectures, indicating an increasing bipolarization of insurance premiums.

Those were the results of analyzing the simplified projects of future insurance premiums for the Greater Tokyo Area. This paper will end by mentioning some tasks for future research.

Although anomalously high indices of increase in premiums by insured persons were uniformly defined as values exceeding 150 throughout the entire duration of this study, statistical methods need to be used to calculate the anomalies. In comparison to Fig. 8.4, the change in the mean insurance premium rates by prefecture should form a smoother curve when anomalies are excluded from the calculations of those mean rates.

Also, due to data availability constraints, a simplified model was used in this study to project the insurance premiums, but using the Ministry of Health, Labor, and Welfare's "Worksheet for Insured Long-Term Care Service Plans" should yield more accurate results.

Furthermore, in this study, insurance premiums were only calculated for the Greater Tokyo Area, but the aging of society is not a phenomenon limited to the Greater Tokyo Area; it affects Japan as a whole, which suggests the need to expand the target region. Doing so will make it possible to use cluster analysis to distinguish regional patterns in changes in insurance premiums.

Finally, we should consider the current nursing care insurance premium policy and then decide whether it is appropriate to estimate future long-term care insurance premiums with constant variables. This is because the current nursing care insurance premium set the amount after decreasing the regional differences as much as possible and therefore we have to assume that the regional differences are constant throughout the future.

References

1. Ministry of Health, Labor, and Welfare. Life Tables for 2016. 2017a. https://www.mhlw.go.jp/toukei/saikin/hw/life/life16/index.html.
2. Inoue T. The Web System of Small Area Population Projection for the Whole Japan (regular version 2.0). 2017. https://www.arcgis.com/1GkdZTX/index.html.
3. Inoue T. Release of a regular version of the web system of small area population projections for the whole Japan. E-J GEO. 2018a;13(1):87–100.
4. Inoue N. GIS analysis of nursing faculties in the Tokyo Metropolitan Area, 2010-2060. Aoyama J Econ. 2018b;70(1):1–10.
5. Yamamoto K. The simulation of the long term care insurance premiums in Japan. In: The 10th IAGG Asia/Oceania Regional Congress, 19-22 October 2015; 2015.
6. Ministry of Health, Labor, and Welfare. Fact-finding survey on project of long-term care for 2017. 2017b. https://www.mhlw.go.jp/topics/kaigo/osirase/jigyo/15/index.html.
7. Ministry of Health, Labor, and Welfare. The first premium and projected service amount of nursing-care insurance in the seventh term planning period. 2018a. https://www.mhlw.go.jp/stf/houdou/0000207410.html.
8. Ministry of Health, Labor, and Welfare. Fact-finding survey on project of long-term care for March 2018. 2018b. https://www.mhlw.go.jp/topics/kaigo/osirase/jigyo/m18/1803.html.
9. Matsuoka H, Nakazawa K. Study on the effect of price revision of long-term care insurance premium on storage rate (Kaigo Hoken Ryo Kakaku Kaitei Ga Shunou Ritsu Ni Ataeru Eikyou). In: RIEB Discussion paper series, DP2017-J05; 2017.

Chapter 9
Advance Care Planning from Clinical Ethics Perspectives in Japan

Kei Takeshita

Abstract Advance care planning (ACP) is a process that aims to provide healthcare in accordance with patients' wishes and values, even after they have lost their decision-making capacity. Government-led ACP initiatives have been underway in Japan since 2018, and the Japanese interpretation of ACP is constantly evolving. This chapter provides an overview of ACP in Japan from the perspective of clinical ethics.

Keywords Advance care planning · Four principles of biomedical ethics · Four topics chart of clinical ethics · Family

9.1 Introduction

In March 2018, Japan's Ministry of Health, Labour, and Welfare (MHLW) revised the "Guidelines on the decision-making process for end-of-life care" (hereafter referred to as "Process Guidelines") [1]. The first edition of the Process Guidelines was developed in 2007, in the wake of a reported incident of ventilator withdrawal at a municipal hospital in Toyama Prefecture in March 2006. The 2018 revision is said to be "based on the need to respond to the development of community-based comprehensive care as an aging and multi-level death society progresses and on the growing prevalence of research and initiatives premised on the concept of ACP (Advance Care Planning), particularly in the western countries [1]."

K. Takeshita (✉)
Department of Medical Ethics, Tokai University School of Medicine,
Shimokasuya, Isehara, Kanagawa Prefecture, Japan
e-mail: takeshita_kei@mac.co

As various definitions of ACP exist and ethical considerations differ depending on the definition used, in this chapter, the author will use the following definition from a leaflet prepared by the MHLW to publicize ACP to the general public published in 2018 [2]: "The ACP (Advance Care Planning) is an effort and process to think in advance about the healthcare you prefer, and to repeatedly discuss and share it with your family members and your healthcare team in case something happens to you." The "something" referred to here is basically defined as "when one loses decision-making capacity [3]."

The purpose of this chapter is to help readers understand ACP from the perspective of clinical ethics. After introducing "Four Principles of Biomedical Ethics" and "Four Topics Chart of Clinical Ethics," the author will focus on "when to use ACP" rather than "when to implement ACP" or "how to implement ACP," and consider the ethics of supporting ACP from the standpoint of clinical ethics while referring to the Process Guidelines [4].

9.2 Four Principles of Biomedical Ethics and Four Topics Chart of Clinical Ethics

9.2.1 Four Principles of Biomedical Ethics

Ethical principles are often used when examining a case from a clinical ethical perspective. A principle in ethics is "a fundamental code of conduct on which many other moral standards and judgments are based [5]." In the Japanese medical community, the four principles of clinical ethics (Respect for Autonomy, Nonmaleficence, Beneficence, and Justice) proposed by Beauchamp and Childress in their textbook "Principles of Biomedical Ethics" [5] are widely accepted (Table 9.1).

The Belmont Report [6] identified three basic principles of research ethics (respect for persons, beneficence, and justice). However, the Principles of Biomedical Ethics have increased the number of principles to four, with the addition of one independent principle, non-maleficence, regarded as a subordinate principle of beneficence. Furthermore, the Belmont Report used the term "respect for persons," while the Principles of Biomedical Ethics used "respect for autonomy."

Since Beauchamp and Childress were colleagues at Georgetown University's Kennedy Institute of Ethics, the use of the Four Principles of Biomedical Ethics to examine ethical issues is sometimes called the "Georgetown Approach [7]." Furthermore, according to Kimura, a well-known Japanese bioethicist, the Four Principles have been described as the "Georgetown mantra" because of "the

Table 9.1 The four principles in "Principles of Biomedical Ethics" [5]

Respect for autonomy
Nonmaleficence
Beneficence
Justice

unintended harm of applying them to any situation [8]." Of the many translations of the Four Principles of Biomedical Ethics available in Japanese, that in the textbook of Akabayashi, a pioneering Japanese physician-ethicist, is often used [9].

9.2.2 The Four Topics Chart of Clinical Ethics

In Japan, when multidisciplinary healthcare teams discuss clinical ethics cases, a quadrant chart ("Four Topics Chart" [10]) is often used to organize the information. The Four-Topic Chart is a two-dimensional graph with the first quadrant representing patient preferences, the second representing medical indications, the third, quality of life (QOL), and the fourth, contextual features. Some facilities have created original case study sheets based on the Four Topics Charts.

By organizing and describing the information in the Four Topics Chart, one can better understand the case and identify missing information that should be examined in more detail. While the Four Principles of Biomedical Ethics tend to be abstract and are considered "top-down" (deductive thinking), the concept of decision-making based on concrete cases is called "bottom-up" (inductive thinking) [11]. The author, who is in charge of clinical ethics consultation as a physician-ethicist at Tokai University Hospital, uses the Four Topic Chart with modifications suitable for our institution to evaluate information in each domain while analyzing it according to the Four Principles of Biomedical Ethics [12].

9.3 "Four Principles of Biomedical Ethics" and ACP

9.3.1 Viewpoint of the Principle of Respect for Autonomy

The principle of respect for autonomy requires healthcare providers to carefully disclose and explain information to patients, explore and confirm patients' level of understanding and spontaneity, and facilitate patients' ability to make appropriate decisions. The following concept in Process Guidelines can be said to be in line with the principle of respect for autonomy: "The most important principle is that the person receiving healthcare should be provided with appropriate information and explanations by physicians and other healthcare professionals and that based on this, the person should have sufficient discussions with the healthcare team consisting of multispecialty healthcare professionals, and that the healthcare at the end of life should proceed based on decision-making by the person [4]."

Here, the author will discuss the role of ACP, focusing on the principle of respect for autonomy in accordance with the classification described in the section of Process Guidelines.

9.3.1.1 Confirmation of the Patient's Wish

When attempting to decide on a medical treatment and care policy, if the patient has sufficient decision-making capacity, the physician informs the patient of the options and their scientific evidence, and the patient then informs the physician of their preferences and wishes. The process of deliberation based on shared evidence and values, aiming to reach a mutually agreed-upon policy, is known as Shared Decision Making ("SDM") and described as the "least bad" physician-patient relationship that balances the principle of respect for autonomy with three other principles: that of non-maleficence, beneficence, and justice [13].

If past ACPs have allowed patients to share their values with the healthcare team and family, and if they have provided patients with sufficient knowledge about the healthcare under discussion, ACPs have greatly benefited the current SDM and informed consent process. Conversely, it would be ideal for ACPs to be implemented in a useful manner.

When a patient loses decision-making capacity, whether the existence of an advance directive created during the ACP process can be considered. In this case, whether or not "the patient's wishes can be confirmed" depends on the relationship between the healthcare team and the patient and the time frame of that relationship. If the healthcare team and the patient have been carefully accumulating ACPs until recently, it may be possible to determine if "the patient's wishes can be confirmed" based on what has been shared with the patient during the ACP process.

9.3.1.2 When the Patient's Wishes Cannot Be Confirmed

The Process Guidelines further distinguish the following three types of cases in which the patient's wishes cannot be confirmed [4]:

1. When family members can presume the patient's wishes, the basic policy is to respect the presumed wishes and the best interests of the patient.
2. When family members cannot presume the patient's wishes, the basic policy is to fully discuss the best interest or wish of the patient with family members who can act on the patient's behalf.
3. When there is no family member or when they leave the decision to the healthcare team, the team should make the healthcare plan consistent with the best interests of the patient.

In any of the three above-mentioned types, there is nothing written in the Process Guidelines about how to actually utilize the ACP. If an ACP was done, and if family members and the current healthcare team were involved in the process, or if the ACP was documented, then the ACP could be utilized to assist the presumption of the patient's wishes. This is in line with the principle of respect for autonomy. It is also conducive to the principle of beneficence if it provides clues in the search for a patient's best interests. It should be remembered that, in the first place, ACP would be implemented because of the expectation that ACP can help with difficult healthcare decisions.

9.3.2 Viewpoint of the Principle of Non-Maleficence

In "Principles of Biomedical Ethics [5]," the chapter "Non-maleficence" includes a section titled "Protecting Incompetent Patients from Harm," in which "Advance Directives" and "Surrogate Decision Making without Advance Directives" are discussed. For example, if the content of the advance directives created during the ACP process or the presumed wishes of the patient by the family members are not considered to be in the patient's current best interest, then one should be cautious in applying them while decision-making.

What should the healthcare team do if the plan shared in the ACP now appears to harm the patient? We might agree to prioritize protecting the patient from harm. If an appropriate ACP process has been repeated and shared fully with family members and the healthcare team, we might conclude that the principle of respect for autonomy is superior to that of non-maleficence. Asking the patient about the extent of the family members' and/or healthcare team's discretion during the ACP process may help the family members and the healthcare team deal with the dilemma between the patient's past decisions and current interests. However, inferring from the fact that even in the Netherlands, where Advance Euthanasia Directives (AEDs) have been institutionalized and legally validated, there is controversy over the application of the same to people with dementia [14], the dilemma would not be completely resolved even if the extent of discretion is defined in such a way. The author believes that it is the responsibility of healthcare providers to confront these dilemmas and arrive at a suitable solution.

The Process Guidelines expect family members who are involved in making healthcare decisions for a patient, to be able to "presume the patient's wishes." If the wishes presumed by family members are inconsistent with the patient's best interests or are considered harmful to the patient, the healthcare plan should be determined through careful discussion, just as if a person with decision-making capacity expressed such wishes. If consensus cannot be reached, the Process Guidelines state that "a separate meeting of multiple experts, including persons outside the healthcare team, should be established to discuss and advise on the healthcare plan," in which case, clinical ethics consultation will likely be requested. Beauchamp and Childress [5] define the "qualification of surrogate decision makers" as (1) the ability to make reasoned judgments (competence), (2) adequate knowledge and information, (3) emotional stability, (4) commitment to the incompetent patient's interests, (5) free of conflicts of interest, and (6) free of controlling influence by those who might not act in the patient's best interests. These points may help select family members and others to presume the patient's wishes.

9.4 "Four Topics Chart of Clinical Ethics" and ACP

9.4.1 Family Members as an Entity to Presume "The Patient's Wishes"

When discussing the healthcare plan of a person with insufficient decision-making capacity, the contents of the records and advance directives that the person talked about during the ACP process are organized as "preferences of patients." However, this is only a convenience to organize the facts, and the onus lies on the healthcare team to decide whether it should actually be adopted as the presumed wishes, referring to the stories of family members.

It is interesting to note that the Process Guidelines read that the expected role of family members is limited to "sharing and communicating the current wishes of the patient toward the future" and "presuming the wishes of the patient who is currently unable to make decisions." For example, even if the patient is currently able to make decisions, the Process Guidelines state, "It is important to have repeated discussions with the patient, the patient's family members and other trusted persons, because the patient may become unable to communicate his or her wishes." The Process Guidelines continue, "It is also important for the patient to designate a certain person from family members or other persons who can presume the patient's wishes in advance of these discussions." Thus, the process guidelines describe the role of the family in terms of "ACP," but do not address what the family should or should not do. It is not assumed that family members will share their own stories.

In situations where the patient is incompetent, family members are expected to communicate the patient's presumed wishes to the healthcare team. If the family members cannot presume the patient's wishes, they are expected to discuss the patient's best interests with the healthcare team. Furthermore, it would benefit the healthcare team if family members could communicate what they knew about the patient's personality and values during the discussion. What family members say in these contexts is basically evaluated as "preferences of patients" in the Four Topics Chart of Clinical Ethics, and is also reflected in "Quality of Life," as discussed below.

9.4.2 Family Members as "Contextual Features"

Healthcare providers tend to focus only on patients (although this is inevitable because they are their clients). Therefore, it is sometimes overlooked that family members may be greatly affected by the healthcare for patients, even though they are independent individuals and have the right to pursue their own well-being. In actual healthcare settings, healthcare planning that serves the patient's own wishes and interests may be welcomed by family members but may also go against the preferences or convenience of family members. Such circumstances of family members are organized as "contextual features" in the Four Topics Chart. Clinical ethics issues sometimes come into sharper focus by examining in detail the information about family members as part of the "contextual features." Healthcare providers need to carefully distinguish whether what family members are saying is "preferences of patients" or are "contextual features." It is important to respect "contextual features" as well as "medical indications," "preferences of patients," and "quality of life" without dismissing them as limiting factors in medical treatment and care to determine a viable healthcare plan.

9.4.3 Consideration of the Relationship Between Patients and Family Members

Even when family members participate in the SDM of a patient with decision-making capacity, there are cases where the patient makes decisions on his or her own, and family members simply share the decision-making process. However, there may be cases where the patient's wishes are formed through discussions with family members, cases where there is a change in the patient's wishes during discussions with family members, and cases where the patient superficially agrees with family members while concealing his or her true wishes. In particular, when it comes to the phase of reflecting the patient's wish in actual healthcare rather than the formation of wishes, it is necessary to reach a consensus with family members who may be physically, financially, and emotionally affected by the implementation of the decision. Therefore, some patients may express their "wishes" to be in line with the intentions of family members in anticipation of the difficulty of reaching a consensus with them. Family members may also change their minds during the discussion process or press the patient to make an acceptable choice. The healthcare team needs to be aware that both the "preferences of patients" and the "contextual features" can be altered by the "discussion" mentioned in the Process Guidelines. However, opinions may differ regarding the extent of involvement of the healthcare team in the dynamics of the patient and family members.

9.4.4 "Contextual Features" Other Than Those of Family Members

The fact that the healthcare team, patients, and family members agree on a healthcare plan does not guarantee that the agreed-upon plan is ethically and legally appropriate. For example, even if everyone agreed on the plan of active euthanasia (death by injection of lethal drugs) if it were actually carried out, the physician who carried out the procedure and the family members who participated in the decision-making process would likely be charged with commissioned murder in Japan. Laws, regulations, including ethics guidelines, and social conventions are essential elements of "contextual features" other than the decisions made by family members.

9.4.5 The Role of ACP in Quality-of-Life Assessment in Incompetent Patients

The "best interest standard," as stated in the Process Guidelines, is the idea that the best healthcare for the patient should be provided when the patient's wishes cannot be presumed. In the "QOL" section of the Four Topics Chart, the patient's best interest is assessed. Jonsen et al. explained that "medical indications" comprise the actions that aim to meet the "needs" of patients, whereas "QOL" refers to the degree of "satisfaction" that people experience and their values [10].

They also noted that the first step in understanding how to apply the concept of QOL is to reflect on the interests that all humans seem to share [10]: (1) an interest in being alive, (2) an interest in being capable of understanding and communicating their thoughts and feelings, (3) an interest in being able to control and direct their lives, (4) an interest in being free from pain and suffering, and (5) an interest in being able to attain desired satisfaction. They then argue that what counts as an interest should be designated as much as possible from the patient's viewpoint.

9.4.6 Cultural Background and ACP

Cultural background is not only one element of the "contextual features" but also influences the coordination of "medical indications," "preferences of patients," "QOL," and "contextual features." Considerations by the healthcare team and clinical ethics consultations are not free from the cultural backgrounds of the patients.

Cheng et al. reviewed ACP in East Asian countries, including Japan, and noted that since most East Asian countries are influenced by Confucianism and the concept of "filial piety," patient autonomy is consequently subordinate to family values and physician authority and that the dominance of family members and physicians during a patient's end-of-life decision-making is recognized as a cultural feature in Asia [15]. It has been noted that it is not only physicians and family members that may inhibit patients' autonomous decision-making; Asai et al. state that healthcare professionals must be aware of the possible adverse effects of psychocultural-social tendencies in Japan, including "surmise," "self-restraint," "air," "peer pressure," and "community," on the implementation of SDM and must promote autonomous decision-making of patients so that they can make choices that sufficiently reflect their individual and personal views of life, experiences, goals, preferences, and values [16]. This appears to be a common thread in ACP practices.

Contrarily, American bioethicist, Sullivan, in a qualitative analysis of interviews conducted with Japanese healthcare professionals during her stay at Kyoto University, discussed that although informed consent in Japan is conceptualized along a spectrum from "jiko kettei"(which is translated as self-determination) to "omakase" (which means to entrust decisions to another person), omakase is not necessarily paternalistic, as it is often initiated by patients themselves [17]. Sullivan also recalls [18]: "What struck me, as an American, most about the informed consent process was the support system for patients and their families." According to the American theory of informed consent, "informed consent is required for a patient's self-determination, and the presence of family members in the room creates the possibility of interfering with the patient's autonomy. In contrast, according to Japanese healthcare providers, the presence of family members in the informed consent process is intended to support the patient's ability to make decisions after receiving an unfavorable diagnosis."

Recently, an attempt to define ACP in Japan was published [19]. They claim to have incorporated the concept of relational autonomy into their definition of ACP because "advance care planning that is based on an individualistic interpretation of autonomy emphasizing self-determination might be difficult to implement in Japan because Japanese older adults often prefer to be vague about complex issues and are often reluctant to express their opinions for fear of becoming a burden on their families." However, it may be somewhat over-optimistic to say that "individuals' capacity for decision-making cannot be determined in a binary presence-or-absence system; it is always supported and enhanced by people close to those individuals."

Akabayashi et al. noted that many patients in Japan may prefer a "family-facilitated approach"(a process of informed consent in which a patient's family communicates with the attending physician and medical staff and often makes treatment-related decisions) and discuss the need to reassess informed-consent styles suitable to the needs of each patient regardless of country he or she resides in, which may indeed be respecting patient autonomy [20]. This is a common concept in ACP.

How should we consider what are perceived as the Japanese cultural characteristics in the informed consent process, SDM, and other discussions about healthcare planning and ACP practice? The author believes that the kind of decision-making process that tends to be preferred as a cultural background and the process of ACP's implementation as a practice and reflecting it in actual healthcare are different issues. Since ACP was born in the Western area, especially in English-speaking countries, it would be difficult to implement it as it is in Japan and other Asian countries. However, if the concept of ACP based on the principle of respect for autonomy is to be significantly modified in Japan due to differences in cultural background, it may be better to reconstruct it as a category separated from ACP efforts.

9.5 Conclusions

The ACP is an initiative supported primarily by the Principle of Respect for Autonomy in the Four Principles of Biomedical Ethics, which is the philosophy underlying Process Guidelines. It is expected to be useful in exploring "preferences of the patient" and "QOL" in the Four Topics Chart of Clinical Ethics when considering medical treatment and care policies for patients with impaired decision-making capacity.

Acknowledgments This chapter contains a discussion written in the book published in Japanese [21] and includes insight from studies on ethics support for home healthcare professionals supporeted by JSPS KAKENHI 19 K10543. The author would like to thank Ms. Tomoko Takano for useful discussions and Editage (www.editage.com) for English language editing.

References

1. Ministry of Health, Labour, and Welfare. Revision of "Guidelines for the Decision-Making Process for Healthcare in the End of Life." (Accessed 14 Mar 2018). https://www.mhlw.go.jp/stf/houdou/0000197665.html. Accessed 27 Oct 2022.
2. Ministry of Health, Labour and Welfare. Dissemination and awareness leaflet for advance care planning. https://www.mhlw.go.jp/content/10802000/000536088.pdf. Accessed 27 Oct 2022.
3. Amano S, Kizawa Y. Opinion on the promotion and awareness of advance care planning. http://zenganren.jp/wp-content/uploads/2019/11/statement_20191129_01.pdf. Accessed 27 Oct 2022.
4. Ministry of Health, Labour and Welfare. Guidelines for the decision-making process for healthcare in the end of life. March, 2018. https://www.mhlw.go.jp/file/04-Houdouhappyou-10802000-Iseikyoku-Shidouka/0000197701.pdf. Accessed 27 Oct 2022.
5. Beauchamp TL, Childress JF. Principles of biomedical ethics. 8th ed. Oxford: Oxford University Press; 2019.
6. The National Commission for the Protection of Human Subjects of Biomedical and Behavioral Research. The Belmont Report. April 18, 1979. https://www.hhs.gov/ohrp/regulations-and-policy/belmont-report/index.html. Accessed 27 Oct 2022.
7. Ainslie DC. Principlism. In: Jennings B, editor. Bioethics, vol. 5. 4th ed. New York: Macmillan; 2014. p. 2485–9.
8. Kimura R. In: Nagayasu Y, Tachiki N, editors. Preface for Japanese version of "principles of biomedical ethics". 3rd ed. Tokyo: Seibundo; 1997.
9. Akabayashi A, Inaba K, Kodama S, et al. Revised version of Introduction to Medical Ethics 1. Tokyo: Keisoshobo; 2017.
10. Jonsen AR, Siegler M, Winslade WJ. Clinical ethics. 9th ed. New York: McGraw-Hill; 2021.
11. Kodama S. Ethical knowledge in clinical ethics. ICU CCU. 2012;36(9):637–42.
12. Dohzono T, Takeshita K, Kamiya K, et al. Handbook of ethics consultation. Tokyo: Ishiyaku Shuppan; 2019.
13. Tilburt JC. Shared decision making. In: Jennings B, editor. Bioethics, vol. 6. 4th ed. New York: Macmillan Reference; 2014. p. 2946–53.
14. Miller DG, Dresser R, Kim SYH. Advance euthanasia directives: a controversial case and its ethical implications. J Med Ethics. 2019;45(2):84–9.
15. Cheng SY, Lin CP, Chan HY, Martina D, Mori M, Kim SH, Ng R. Advance care planning in Asian culture. Jpn J Clin Oncol. 2020;50(9):976–89.
16. Asai A, Okita T, Bito S. Discussions on present Japanese Psychocultural-social tendencies as obstacles to clinical shared decision-making in Japan. Asian Bioeth Rev. 2022;14(2):133–50.
17. Specker SL. Dynamic axes of informed consent in Japan. Soc Sci Med. 2017;174:159–68.
18. Sullivan LS. Informed consent and support for patients and families in Japan. Kokoro no Mirai. 2015;14:40–1. http://kokoro.kyoto-u.ac.jp/jp/kokoronomirai/kokoro_vol14.pdf. Accessed 27 Oct 2022.
19. Miyashita J, Shimizu S, Shiraishi R, et al. Culturally adapted consensus definition and action guideline: Japan's advance care planning. J Pain Symptom Manage. 2022;64:602.
20. Akabayashi A, Slingsby BT. Informed consent revisited: Japan and the U.S. Am J Bioeth. 2006;6(1):9–14.
21. Takeshita K. ACP wo sapo-to suru rinri (ethics in support of ACP). In: Kizawa Y, Shima Y, Takamiya Y, Tsuneto S, Miyshita M, editors. Hosupisu kanwa kea hakusho 2023(hospice palliative care white paper 2023). Tokyo: Seikaisha; 2023. p. 16–22.

Chapter 10
Social Capital Well-Being in a Super-Aging Society

Masumi Takada and Nobuko Miyata

Abstract In the twenty-first century, the "Health Japan 21" initiative for the second phase of national health promotion focuses on the ultimate goal of "extending healthy life expectancy and reducing health disparities." This initiative emphasizes not only individual health support but also the creation of a societal environment where the entire community collaboratively safeguards health. The utilization of social capital is highlighted as a means to showcase the impact on the health and well-being of the elderly population.

Keywords Social capital · Community well-being · Aging society · Health disparities · Integrated care system

10.1 Introduction

Amidst the rapid economic growth following World War II, Japan achieved the world's highest life expectancy through nutrition improvements and healthcare technology advancements. While this achievement is undoubtedly a source of pride, it also underscores various challenges stemming from societal factors and living environments.

By the year 2025, it is estimated that the population of individuals aged 75 and above will comprise 18.1% of the total, and by 2040, the population over 65 years

M. Takada (✉) · N. Miyata
Faculty of Nursing and Rehabilitation, Chubu Gakuin University, Gifu, Japan
e-mail: takada-masumi@chubu-gu.ac.jp

old will account for about 35% of the total population [1]. For the fiscal year 2022, the average life expectancy stood at 81.05 years for males and 87.09 years for females [2]. The average number of years individuals can live without restrictions on their daily lives (healthy life expectancy) is reported as 70.4 years for males and 73.6 years for females. Consequently, a significant portion of those aged 75 and above are expected to require some form of medical or caregiving assistance [3]. Particularly in regions with lower social capital (SC), there is an emphasis on enabling individuals to "live in their accustomed homes and communities as themselves" for as long as possible, necessitating support at the local government level.

To ensure the quality of life (QOL) for elderly members of communities, it is imperative to establish an integrated framework encompassing housing, medical care, caregiving, prevention, and lifestyle assistance. Acknowledging this, the Ministry of Health, Labour and Welfare has been promoting the "Community-Based Integrated Care System" since 2003. While coordination with specialized institutions in healthcare, medical care, and caregiving is pivotal in operating the Community-Based Integrated Care System, there is also growing interest in leveraging social capital for community activities.

10.2 Concept of Social Capital

The concept of social capital traces its origins back to novelist Henry James and subsequently gained scholarly attention in fields such as education, sociology, and economics. Robert Putnam further propagated the notion of social capital across disciplines, including political science, economics, sociology, social psychology, management, education, and public health. Putnam defines social capital as "features of social organization, such as trust, norms, and networks, that can improve the efficiency of society by facilitating vigorous cooperation among people engaged in various coordinated activities, leading effectively to cooperative action toward common goals." [4].

Trust
Putnam differentiates between "thick trust (assets of close social networks)" and "thin trust (general trust in other members of the community)," highlighting that "thin trust" promotes broader cooperative behaviors. Mutual trust fosters spontaneous cooperation, and such cooperation, in turn, nurtures trust. Putnam thus perceives trust as both an essential component of social capital and something that social capital generates.

Norms
Among various norms, Putnam particularly underscores the norm of reciprocity. Reciprocity involves interdependent exchange of benefits, categorized into balanced reciprocity (simultaneous exchange of equally valued items) and generalized reciprocity (sustained exchange based on mutual expectations for future balance,

even if initially unequal). Generalized reciprocity draws from altruism in seeking immediate benefits for the other party and from self-interest in enhancing utility for all parties in the long term, contributing to the harmony of self-interest and solidarity.

Networks

Networks encompass vertical networks like employer-employee relations within workplaces and horizontal networks such as choirs and cooperatives. Putnam's Italian study suggests that densely-knit vertical networks cannot sustain social trust and cooperation to the same extent as horizontally-knit networks driven by active civic participation, like neighborhood groups or sports clubs. In essence, the development of "horizontal networks" in northern Italian regions and "vertical networks" in southern states is highlighted. A broad emphasis is placed on "weak ties" beyond immediate family, particularly valuing "face-to-face networks" as a core element.

Regarding the relationship between these three elements of social capital, Putnam indicates the possibility of social trust arising from the norms of reciprocity and active civic participation in networks. He further argues that an increase in any one element leads to an increase in others, suggesting mutual reinforcement.

10.3 Social Capital and Health

The positive impact of fostering social capital (SC) on health has been demonstrated in various studies [5–7]. Research on individuals' subjective health perceptions suggests that lower levels of SC elements are associated with lower subjective health perceptions in men and an inclination towards promoting depression in women [8]. Studies related to caregiving reveal that women in areas with lower levels of community trust have a 68% higher risk of becoming dependent on care compared to women in areas with higher trust. Furthermore, encouraging participation in community activities has been shown to contribute to frailty prevention. SC elements like reciprocity and norms enhance trust in the community, and community participation promotes mutual acquaintanceship among residents. This fosters the creation of trustworthy companionships and communal spaces, impacting not only frailty prevention but also subjective health perceptions [9].

The cultivation of SC does not solely depend on the individual level; rather, it is significantly influenced by the physical, human, and social environments of the local and neighboring communities. In particular, interactions among residents and engagement with neighbors are essential elements facilitating proactive initiatives to address health issues and their resolutions. In contemporary Japan, maintaining connections with neighbors without departing from one's childhood community might be considered uncommon. Given this context, it is crucial to support the cultivation of SC not only among residents but also through the utilization of the functions of relevant institutions, including governmental agencies, to ensure the facilitation of SC in the community.

10.4 Elderly Living

In the case of elderly individuals, there is a demand for support not only in caregiving but also in their daily lives. However, there are instances where certain cases fall outside the scope of long-term care insurance coverage, making it challenging for the government alone to provide comprehensive support for residents' lives. Particularly in areas facing depopulation, the scarcity of support for daily living and caregivers exacerbates the difficulty of solely relying on long-term care insurance for preventive care and resident support. Consequently, collaborative efforts and coordination between residents and the government, such as community activities and volunteer initiatives, are imperative, particularly in regions deeply affected by population decline.

10.5 Investigative Research on Elderly Residents' Sense of Attachment to the Community ~Focusing on Elderly Residents in Depopulated Mountainous Areas ~

10.5.1 Purpose

In various municipalities, efforts are actively being made to support the medical, welfare, and caregiving needs of elderly residents who live and recuperate within the community, with a central focus on Comprehensive Community Support Centers. The establishment of networks among community residents is a vital aspect of promoting "Comprehensive Community Care." In the surveyed region, clear strategies have been outlined for healthcare, medical, and welfare collaboration through avenues such as home-based end-of-life care and dementia support. Amidst such initiatives, this study aimed to investigate and explore the relationship between mutual support among community residents and their attachment to the local area as pertinent factors contributing to the continuity of living in familiar surroundings.

10.5.2 Method

A questionnaire-based survey was conducted between October and November 2016, targeting approximately 900 households of elderly residents aged 65 and above in a mountainous depopulated area with an aging rate of 43.2%. The survey distinguished three groups: (1) elderly individuals aged 65 and above living with family members (general elderly), (2) elderly individuals aged 65 and above living alone (solitary elderly), and (3) those deemed likely to require support or caregiving in the near future (specific elderly). The questionnaire covered basic information such as gender, current health status, issues in daily life and their resolution methods, mutual assistance among residents, and levels of attachment to the community. Chi-square tests were conducted for each question item to analyze the relationship between mutual support among elderly community residents and their attachment to the local area.

10.5.3 Ethical Considerations and Conflicts of Interest

Prior to the survey, participants were provided explanations about the purpose of the study, the anonymity of survey responses, and the absence of adverse effects for refusing to participate. There were no identified conflicts of interest that required disclosure related to the content of the presentation or the presenters.

10.5.4 Result

We conducted a survey targeting a total of 925 households, from which we obtained responses from 721 households. Using a chi-square test, we analyzed the relationship between various factors among the surveyed elderly population, including their current health status, the presence of daily life challenges, methods of resolving these challenges (whether they seek help from others), and their willingness to contribute to others.

We found correlations between the presence of problem-solving methods for challenges and the desire to continue living in the community in the future, as well as the degree of attachment to the local community ($p < 0.05$). Among those who indicated the existence of methods to address their challenges (seeking help from others), we observed a significant positive correlation with satisfaction with the local community (Table 10.1). Similarly, individuals who believed they could contribute to others were significantly more inclined to want to continue living in the local area (Table 10.2). Furthermore, there were no significant differences in terms of health status and attachment to the community between the general elderly population and specific elderly subgroups.

Table 10.1 About whether or not there is a solution and whether or not they want to continue living in the village

		Solution			
		No answer	Absence	Presence	Total
Problem	No answer	7	1	7	15
	Absence	81	29	88	198
	Presence	19	12	59	90
	Total	107	42	154	303

$\chi^2 = 13.59 > \chi_0 = 9.49$

Table 10.2 The extent to which one desires to support residents of their community in continuing to reside in the current area

Consider helping others		To continue living in the village → negative						
		1	2	3	4	5	6	Total
	No answer	5	32	21	4	0	1	63
	Absence	2	11	6	3	0	1	23
	Presence	5	125	69	15	2	1	217
	Total	12	168	95	22	2	3	303

$\chi^2 = 216.74 > \chi_0 = 18.31$

10.5.5 Discussion

The study explored the relationship between health status and attachment to the community. Though no significant correlation was found, it revealed that the presence of available problem-solving methods was associated with higher community satisfaction. Additionally, respondents who believed they could contribute to others exhibited significantly higher interest in continuing to reside in the community. This implies that mutual support among local residents enhances their attachment to the area. The surveyed region had a history of emphasizing human networks within the community, such as implementing dementia support initiatives, even before the comprehensive community care system gained full momentum. Such initiatives likely contribute to the robust connections among community residents. Kawachi et al. have reported that "trust," "reciprocal relationships," and "connections with society" deeply influence longevity. In this study, it was observed that even specific elderly individuals with health risks showed a propensity for strong attachment to the local community. Additionally, the substantial number of respondents indicating their ability to assist others could potentially bolster the development of a comprehensive human network within the community.

References

1. Ministry of Health, Labour and Welfare, Population Trends in Japan. https://www.mhlw.go.jp/seisakunitsuite/bunya/hukushi_kaigo/kaigo_koureisha/chiiki-houkatsu/dl/link1-1.pdf. Accessed 18 Sept 2023.
2. Ministry of Health, Labour and Welfare, Summary of the 2022 Simplified Life Chart. https://www.mhlw.go.jp/toukei/saikin/hw/life/life22/dl/life22-02.pdf. Accessed 18 Sept 2023.
3. Cabinet Office, Trends in Healthy Life Expectancy. https://www8.cao.go.jp/kourei/whitepaper/w-2022/zenbun/pdf/1s2s_02.pdf. Accessed 18 Sept 2023.
4. Putnam RD. Making democracy work: civic traditions in modern Italy. Princeton, NJ: Princeton University Press; 1993.
5. Rose R. How much does social capital add to individual health? A survey study of Rossians. Soc Sci Med. 2000;51(9):1421–35.
6. Pollack CE, von dem Knesebeck O. Social capital and health among the aged: comparisons between the United States and Germany. Health Place. 2004;10(4):383–91.

7. Forsman AK, Nyqvist F, Wahlbeck K. Cognitive components of social capital and mental health status among older adults: a population-based cross-sectional study. Scand J Public Health. 2011;39(7):757–65.
8. Ohta H. Associations between individual-level social capital and self-rated health or depression among elderly men and women. Jpn J Public Health. 2014;61(2):71–85.
9. Yoshiyuki N. Association between fraily in community-dwelling older adults certified as requiring support in the long-term care insurance system and social capital among local neighborhood volunteers. Jpn J Public Health. 2020;67(2):111–20.

Chapter 11
Development of Medical Technology for Social Isolation in an Aging Society Through an Industry–Government–Academia Collaboration: Okumikawa Medical Valley Project

Hidemasa Yoneda, Ryohei Hasegawa, and Hitoshi Hirata

Abstract To solve the social isolation of the elderly requires a mechanism; it is necessary to promote the social participation of the elderly through an interdisciplinary approach that includes the humanities and social sciences. We have created a co-create platform Okumikawa Medical Valley, which collaborates with industry, government, and academia to utilize science and technology to explore the image of a community with a high level of well-being. This chapter introduces the field demonstrating aging countermeasures such as presenteeism surveys, remote rehabilitation and ultrasound treatment using high-speed lines, and a cognitive function survey using sign language recognition.

Keywords Social isolation of the elderly · Collaboration with industry, government, and academia · Presenteeism surveys · Remote rehabilitation and ultrasound treatment · Cognitive function survey

---

H. Yoneda (✉) · H. Hirata
Human Enhancement and Hand Surgery, Nagoya University, Nagoya, Japan

R. Hasegawa
Human Enhancement and Hand Surgery, Nagoya University, Nagoya, Japan

National Institute of Advanced Industrial Science and Technology, Tsukuba, Japan

11.1 Introduction

Social isolation of the elderly is caused by the loss of local communities because of urbanization, a shift to nuclear families, and a decline in the marriage rate. It has become a serious social problem in an aging society. Japan's population is aging rapidly, and the percentage of the elderly aged 65 years and over in 2020 will be the highest in the world, at 28.6%. At the same time, 24.2% of households in Japan, where the population is 65 years or older, live alone; this is the highest percentage worldwide [1]. This percentage is estimated to reach 40% by 2040. The problem of an aging society and its accompanying social isolation is accelerating, but solutions to this problem are yet to be found.

Several reports have shown that social isolation among older adults decreases happiness. Waldinger found that contributions to society and human connections are sources of happiness [2]. Matsuura et al. reported that the level of happiness decreases for elderly people living alone, especially among men [3]. Many Japanese people experience happiness by continuing to work after retirement. Therefore, even if physical decline due to aging is inevitable, people can live happily if provided with opportunities to continue their social participation [4].

One of the reasons for why elderly people are hesitant to participate in society and continue working is their perceived decline in brain function. However, even if brain atrophy is detected by brain function imaging, crystalline intelligence is maintained at a high level [5]. Therefore, maintaining fluid intelligence, which declines with age, can remove barriers to social participation and reemployment after retirement. Various brain training programs have been proposed to maintain and train the fluid intelligence of the brain; however, these are limited to those performed as individuals. Available brain training programs do not include maintaining and improving brain functions through interaction with others and social participation; therefore, they are incomplete tools for social isolation.

Furthermore, social isolation cannot be solved simply by introducing new technologies; it requires a multidisciplinary approach from the humanities and social sciences to promote the social participation of the elderly. Therefore, it is necessary to create a platform where industry, government, and academia can co-create, utilize innovative science and technology, and search for an image of a community with a high level of well-being, while repeatedly conducting proof-of-concept for this ideal. This chapter introduces the work conducted thus far in the Okumikawa Medical Valley (OMV), a demonstration field for aging countermeasures, using the framework of industry–government–academia collaboration that the authors are pioneering [6].

11.2 Overview of Okumikawa Medical Valley

Okumikawa, located in the northeastern part of Aichi Prefecture, is a mountainous natural region where agriculture and forestry have thrived for centuries [7]. It is an important region in the history of Japan. The Nagashino War occurred in 1575, which led the Tokugawa and Oda families to seize political power. In recent years, young people have moved to neighboring urban areas, resulting in rapid aging of the population, and the area has become an aged, depopulated area, with the percentage of the elderly over 65 years of age being approximately 40%.

Nagoya University has predicted that the Okumikawa region would be an ideal field for demonstrating new technologies related to aging from the perspective of preventive medicine. Under a comprehensive agreement with Shinshiro City, Nagoya University has conducted various research and development activities, such as the OMV, and held events and exchanges with citizens [8]. The OMV has been conducting a variety of research and development activities, as well as events for exchanges with citizens.

In particular, we organized the "Tsukude Charette"in the Tsukude area of Shinshiro City, an isolated mountainous community with 2500 residents. This was a co-creation consortium consisting of representatives of residents, local government officers, city hall staff, the Board of Education, and council members. In the following sections, we introduce our activities to address the social isolation of the elderly in the Tsukude Charette.

11.3 Assessing the Proactive Productivity of the Elderly Through Presenteeism Surveys

The first effort in OMV was to address the issue of detecting people who are physically or mentally impaired or at high risk of such impairments due to social isolation. The authors used the Presenteeism Survey, a healthcare indicator developed by the World Health Organization, to address this challenge [9, 10]. Presenteeism has been shown to be capable of assessing the proactive productivity of workers, but also of housewives and retirees in terms of hobbies and daily activities [11]. In this project, we developed an online system to assess the proactive productivity of the elderly. The system will be implemented as a questionnaire-based health survey targeting residents. It is expected to identify people with physical problems or health concerns, leading to interventions such as early visits to medical institutions. Health risks can be detected with high sensitivity by conducting a pulse survey with repeated periodic assessments and using the range of variation as an indicator. Moreover, if operated over a wide area, people with health concerns can be detected and nudged to change their behavior at the regional and individual levels.

However, many elderly people are likely to face a digital divide (information gap), and it is anticipated that it will be difficult to use mobile terminals as input devices. Tsukude has a poor reception environment. However, almost all households have a contract with a local cable TV company, and their TV sets are connected by an optical fiber line. Therefore, we constructed a system that enables pulse surveys using a remote TV and controller, which enables us to evaluate the health status of many residents. In the future, the government's involvement in the operation of the system will enable a longitudinal analysis of social isolation, and we are currently working on this project with a cable TV provider.

11.4 Remote Rehabilitation Supervision and Ultrasound Examination Using High-Speed Data Communication for Patients Who Have Difficulty Visiting a Hospital (Fifth-Generation Mobile Network)

The use of high-speed data communication (5G) has been increasing as a solution for patients who have difficulty visiting hospitals in remote areas like mountainous regions with depopulation, low birthrates, and an aging population. We conducted a proof-of-concept study of remote medical care and rehabilitation guidance for patients in hospitals and ultrasonography for disaster medical care in a 5G environment [12].

Physical therapists who provide rehabilitation instructions remotely need to assess the patient's general condition and local movements of limbs in greater detail than those who provide rehabilitation on-site. Therefore, an environment in which judgments can be made based on high-resolution images acquired from multiple directions is necessary. Hence, we investigated whether physiotherapists can remotely instruct patients using high-resolution images from multiple directions, such as gait training, when conducting rehabilitation instructions. After interviewing the subjects remotely, various vital parameters such as blood pressure, body temperature, pulse, SpO_2, and electrocardiogram were collected. During subsequent rehabilitation, both the overall image from the fixed 4 K camera and local images from the nurse's head-mounted camera were transmitted to the remote physiotherapist in real-time (Fig. 11.1). Motion capture system images were also transmitted during gait training. The speed, stride length, and joint angles during gait were measured (manuscript in preparation?). The subjects responded that rehabilitation could be performed with peace of mind, but the healthcare providers who participated in the demonstration expressed anxiety owing to the inability to recognize

Fig. 11.1 Remote rehabilitation using a high-speed data line. The physiotherapist's instructions were transmitted to the subject through the monitor to demonstrate rehabilitation

Fig. 11.2 Remote ultrasound examination. The evaluating physician supervises the ultrasound examinations for the diagnosis by sending instructions to the healthcare provider who is in contact with the subject

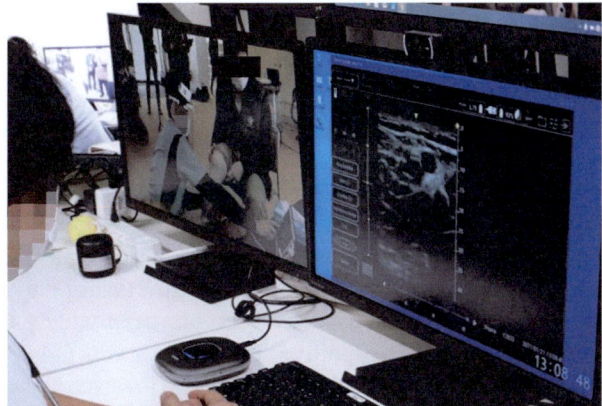

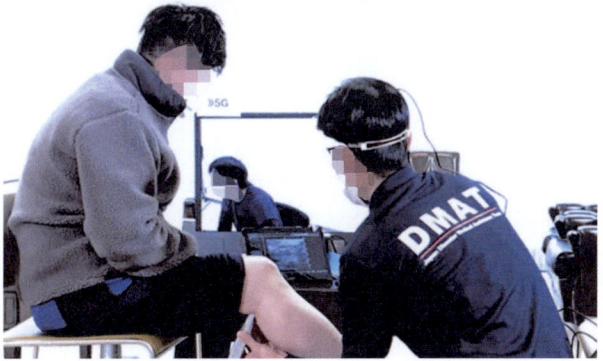

subtle changes in the subjects' facial expressions and the difficulty in communicating with the subjects through the screen.

In addition, in the case of disaster medical care, mountainous villages may become isolated because of blocked roads during disasters or snowfall. In the event of a disaster, the risk of developing deep vein thrombosis, arrhythmia, hypertension, fever, and pneumonia is high and early treatment is necessary. Physicians should be able to detect sudden changes in residents' symptoms remotely and determine whether emergency helicopter transport is necessary or can be handled on-site. We demonstrated remote medical treatment for deep vein thrombosis, which is a problem in times of disaster, by including the transmission of ultrasound images of the lower extremities (Fig. 11.2). Images from a head-mounted camera worn by a local nurse were transmitted to the hospital in real-time, and a remote physician provided instructions on probe manipulation. Consequently, image transmission was possible (manuscript in preparation?). However, there was an image delay due to the frame rate problem.

11.5 Cognitive Function Evaluation Using Sign Language Recognition

Frailty is a state of increased vulnerability to health disorders due to various functional changes and decreased reserve capacity associated with aging [13]. Frailty is associated with a higher risk of adverse health outcomes, including falls, hospitalization, disability, and mortality. Identifying frail individuals allows healthcare professionals to implement appropriate interventions and preventive measures to improve their health outcomes. Although the original concept of frailty was developed from a physical aspect, which has been widely used worldwide, it has been expanded to include cognitive and social aspects [14–16]. It is, therefore, important to evaluate frailty from these functions. However, these were performed separately in the existing efforts, and it was difficult to comprehensively evaluate the subjects. Therefore, we focused on "sign language."

Sign language is widely used as a communication tool by deaf individuals. In the process of learning sign language, in addition to the cognitive function of imitating the shape and movement of other people's hands [17], it is also necessary to have the motor function of reproducing the shape and movement using one's own hands. Social functions are necessary for communicating with others using sign language. Therefore, we developed a cognitive training system using sign language to detect frailty early, including cognitive, motor, and social functions.

First, we conducted experiments to measure quantitative behavioral indicators and brain activity related to sign language recognition. To simplify the process for beginners of sign language, we first used eight photographic images of "finger letters" that can be expressed as still images (Fig. 11.3). Then, based on our previous

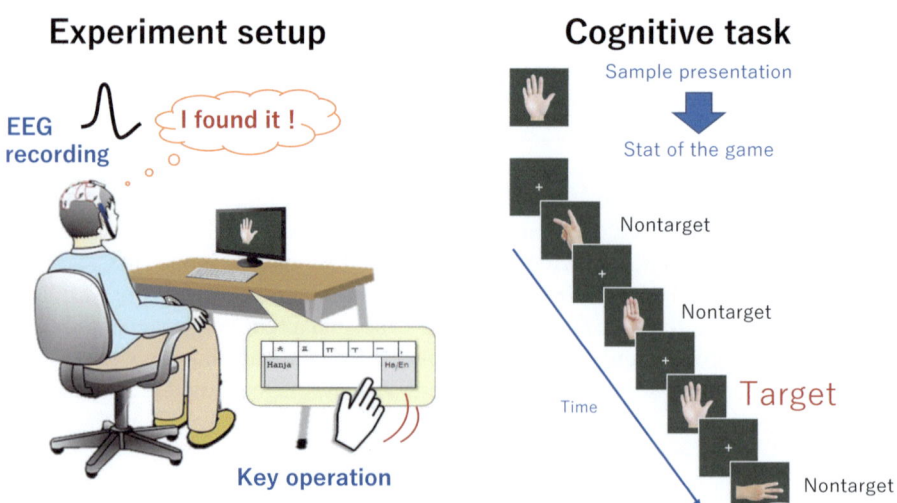

Fig. 11.3 Cognitive task to evaluate the frailty using sign language

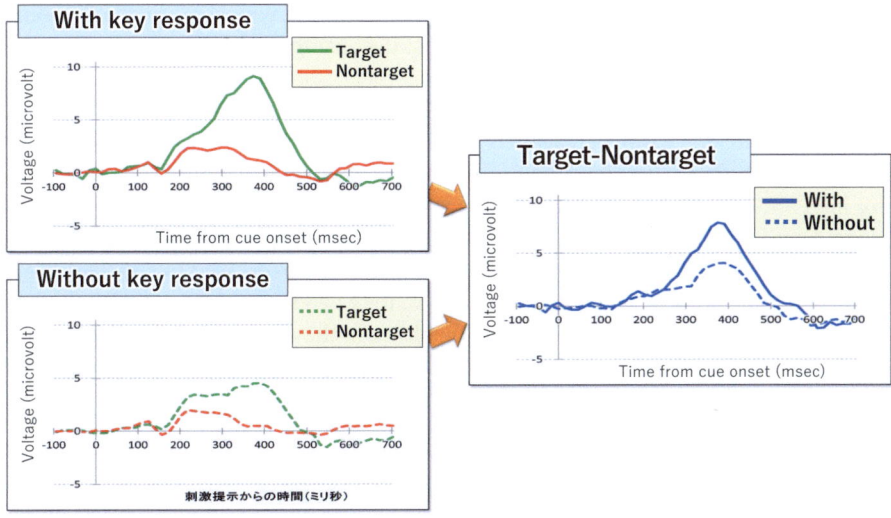

Fig. 11.4 Event-related potential to target or nontarget with/without key response

research using simple figs [18]., we developed a computer-controlled system that presents these images at 1-s intervals, like flash cards [19].

To evaluate the performance of this system, exploratory experiments were first conducted on ten healthy non-elderly adults. After wearing headgear for electroencephalogram (EEG) measurements, eight types of finger letter images were presented pseudo-randomly. The participants were asked to respond by operating the keyboard when they were presented with the same image as the previously taught finger letter (the target).

As a result of the behavioral experiment, the average correct answer rate was over 90% and the average response time was within 0.5 s. In addition, EEG measurements revealed stronger potential changes (event-related potentials reflecting temporal changes in attention [20, 21]) for the target finger letter than for the other finger letters. The results also showed that the response was stronger under the keyboard operation condition, which may require more concentration (Fig. 11.4).

These results indicate that recognition of finger letters is moderately difficult in cognitive training because many healthy subjects can respond quickly and accurately and that the degree of difficulty for individual subjects can be visualized as changes in brain activity. In the future, we will expand the stimulus set from still images (finger letters) to videos (sign language actions at the word level) and expand the target users to include the elderly. We plan to examine whether any individual has suspected frailty. We also plan to examine the associations with other cognitive tests and social situations when frailty is suspected.

The experiments described thus far are mainly designed to evaluate cognitive functions using sign language; however, even if a person can accurately recognize the presented fingers and sign language actions, whether they can reproduce them using their own fingers is another matter. We considered a quantitative evaluation using an image analysis system for hand movements, which is currently being developed separately. Social interaction skills using sign language must be addressed.

Thus far, we have shown that our cognitive training system using sign language may be useful for comprehensively evaluating frailty in elderly populations. However, we believe that repeated use of this system may also contribute to preventive interventions for frailty. Thus, we focused on sign language songs. A sign language song is an activity that resembles dance choreography in matching the song because lyrics that are originally expressed by voice are expressed in sign language. When players memorize sign language words, they can associate them with the lyrics of a song, which aids their memory. In addition, they enjoyed the cognitive task in a game-like manner. Furthermore, if multiple people can "chorus" the sign language song at the same time, it can be expected to deepen their sense of solidarity and provide opportunities for social interaction.

11.6 Summary

This chapter introduces various trials and developments related to the elderly conducted in the Okumikawa Medical Volley Project. Here, research utilizing information technology such as high-speed communication and cutting-edge neurotechnology is conducted to solve social isolation.

References

1. Research NIoPaSS. Estimated future number of households in Japan 2019. https://www.ipss.go.jp/pp-pjsetai/j/hpjp2019/t-page.asp. Accessed 18 June 2023. In Japanese.
2. Waldinger RJ, Schulz MS. What's love got to do with it? Social functioning, perceived health, and daily happiness in married octogenarians. Psychol Aging. 2010;25(2):422–31.
3. Matsuura T, Ma X. Living arrangements and subjective Well-being of the elderly in China and Japan. J Happiness Stud. 2022;23(3):903–48.
4. Japan Cabinet Office. Aging population. 2023. https://www8.cao.go.jp/kourei/whitepaper/index-w.html. Accessed 18 June 2023. In Japanese.
5. Schaie KW. Developmental influences on adult intelligence: the Seattle longitudinal study. 2nd ed. New York: Oxford University Press; 2013. viii, 587-viii.
6. Okumikawa Medical Valley Project. https://omv.aichi.jp. Accessed 18 June 2023.
7. Collaborations on Medicine, health and lifestyle. https://www.city.shinshiro.lg.jp/sangyo/sangyo/jisedai/houkatsurenkei.html. Accessed 18 June 2023. In Japanese.
8. Youtube Channel: Okumikawa Medical Valley Project. https://www.youtube.com/channel/UC8zRRN5B1fJqobXkQJLlLvQ. Accessed 18 June 2023.
9. Kessler RC, Ames M, Hymel PA, Loeppke R, McKenas DK, Richling DE, et al. Using the World Health Organization health and work performance questionnaire (HPQ) to evaluate the indirect workplace costs of illness. J Occup Environ Med. 2004;46(6 Suppl):S23–37.
10. World Health Organization. Health and Work Performance Questionnaire. https://www.hcp.med.harvard.edu/hpq/. Accessed 18 June 2023.

11. Suzuki T, Miyaki K, Sasaki Y, Song Y, Tsutsumi A, Kawakami N, et al. Optimal cutoff values of WHO-HPQ presenteeism scores by ROC analysis for preventing mental sickness absence in Japanese prospective cohort. PLoS One. 2014;9(10):e111191.
12. Saeki M, Oyama S, Yoneda H, Shimoda S, Agata T, Handa Y, et al. Demonstration experiment of telemedicine using ultrasonography and telerehabilitation with 5G communication system in aging and depopulated mountainous area. Digit Health. 2022;8:20552076221129074.
13. Fried LP, Tangen CM, Walston J, Newman AB, Hirsch C, Gottdiener J, et al. Frailty in older adults: evidence for a phenotype. J Gerontol A Biol Sci Med Sci. 2001;56(3):M146–56.
14. Walston J, Hadley EC, Ferrucci L, Guralnik JM, Newman AB, Studenski SA, Ershler WB, Harris T, Fried LP. Research agenda for frailty in older adults: toward a better understanding of physiology and etiology: summary from the American Geriatrics Society/National Institute on Aging research conference on frailty in older adults. J Am Geriatr Soc. 2006;54(6):991–1001.
15. Rockwood K, Stadnyk K, MacKnight C, McDowell I, Hébert R, Hogan DB. A brief clinical instrument to classify frailty in elderly people. Lancet. 1999;353(9148):205–6.
16. Sutton JL, Gould RL, Daley S, Coulson MC, Ward EV, Butler AM, et al. Psychometric properties of multicomponent tools designed to assess frailty in older adults: a systematic review. BMC Geriatr. 2016;16:55.
17. Iacoboni M, Woods RP, Brass M, Bekkering H, Mazziotta JC, Rizzolatti G. Cortical mechanisms of human imitation. Science. 1999;286(5449):2526–8.
18. Hasegawa R, Nakamura Y. Properties of the event-related potentials during the target selection task: toward the development of the cognitive assessment system. Transact Japan Soc Kansei Eng. 2020;19(1):89–96. [In Japanese].
19. Hasegawa RP, Yoneda H, Iwatsuki K, Oyama S, Saeki M, Yamamoto M, Hirata H. Characteristics of event-related potentials during a target selection task with fingerspelling images: using a quantitate evaluation by an EEG-based BMI. Transact Japan Soc Kansei Eng. 2023;22(1):21–9. [In Japanese].
20. Sutton S, Braren M, Zubin J, John E. Evoked-potential correlates of stimulus uncertainty. Science. 1965;150(3700):1187–8.
21. Polich J. EEG and ERP assessment of normal aging. Electroencephalogr Clin Neurophysiol. 1997;104(3):244–56.

Chapter 12
The Diffusion Process of Spanish Influenza Deaths in Japan: Mathematical Models Using the Gompertz and Logistic Curves

Takashi Inoue

Abstract This chapter examines the diffusion process of deaths due to the Spanish influenza pandemic using mortality statistics from Japan. For this purpose, the study constructs four modified models, based on the Gompertz and logistic curves, that incorporate an initial effect and a winter effect. These models are then applied to the second wave of the Spanish influenza. In applying them, Japanese prefectures are classified into five types; the average monthly mortality rates are calculated for each type, and the calculated cumulative mortality rates are used as the models' variables. The results of the model analysis can be summarized as follows: First, the winter effect appears to be stronger than the initial effect. Second, the Gompertz curve-based model tends to better fit region types with relatively high population density, while the logistic curve-based model tends to better fit the opposite types. Third, region types following the Gompertz curve typically have higher final cumulative mortality rates than those following the logistic curve. Since the high influenza mortality rate among the elderly is a significant issue in contemporary Japan, these results can be expected to contribute to improving that mortality rate and advancing gerontology.

Keywords Spanish influenza · Diffusion process · Mortality · Gompertz curve · Logistic curve · Japan

T. Inoue (✉)
Department of Public and Regional Economics, Aoyama Gakuin University, College of Economics, Tokyo, Japan
e-mail: t-inoue@cc.aoyama.ac.jp

12.1 Introduction

The declining trend in the mortality rate of Japan over several decades before World War II is often understood based on demographic transition; however, the increase in this rate during the 1920s is contrary to this trend. This reversal must have originated in the Spanish influenza pandemic, spreading across Japan and the world from 1918 to 1920. This flu virus is typically called "Spanish cold" in Japan; however, "cold" is generally caused by various viruses infecting the respiratory tract [1]. Accordingly, it is inappropriate to refer to this flu as a "Spanish cold." Therefore, following Hayami [2], this study refers to it as the Spanish influenza or simply the Spanish flu.[1]

The first documented case of Spanish flu is believed to have occurred in March 1918 at a military camp in Kansas, the United States [3]. However, the full-scale epidemic began in the spring and summer of the same year, spreading among soldiers on the Western Front in France [4]. Approximately 600 million people worldwide became infected with the Spanish flu, resulting in at least 25 million deaths. In Japan, according to "Ryukosei kanbo" ("Epidemic cold") published by the Health Bureau of the Ministry of Home Affairs [5], approximately 23.58 million people were infected, and more than 385,000 died. These estimates are obviously enormous, considering that the worldwide population then was approximately 1.8 billion and that of mainland Japan was approximately 55 million. The death toll from the Spanish flu was the most adverse in human history among single infectious diseases and other natural or artificial disasters [2]. Hayami [2] provided a comprehensive and detailed discussion of the historical realities of the Spanish flu in Japan and globally.[2]

The Spanish flu caused enormous suffering in Japan. However, there is a limited body of literature that focuses on the Spanish flu in Japan, aside from works related to medicine and the studies of Sugiura [4], Ikeda et al. [7], and Hayami [2]. Of these, only Sugiura's [4] study discussed the Spanish flu epidemic in Japan using mathematical and statistical methods. He calculated the mortality rate from the Spanish flu by prefecture using the "Japan Imperial Death Cause Statistics" (hereafter, JIDCS) published by the Bureau of Statistics [8–10]. Then, by applying factor analysis to the calculated rates, he estimated how the flu diffused spatially, thus extracting its endemic area. This analysis found that the Spanish flu entered the country primarily from major ports in western Japan, spreading throughout the country and, possibly, among small and middle cities via Tokyo from Yokohama Port. Although Ikeda et al.'s [7] report was published in a medical journal, it explored the spread of Spanish flu, presenting many distribution maps of mortality rates by prefecture. However, no mathematical statistical analysis of those findings has been performed. Furthermore, Sugiura's [4] study used factor analysis only as a statistical method, and no verification was performed using a mathematical model of epidemic spread

---

[1] According to Hayami [2], although the influenza epidemic from 1918 to 1920 had no significant association with Spain, the word "Spanish" became firmly associated with this flu, leading to the term "Spanish flu."

[2] Okada [6] referred to the Spanish flu, quoting Hayami [2].

and transmission. Therefore, this study applies a mathematical model to Japan's Spanish flu death statistics and uses it to examine the diffusion process of deaths from the flu.

As in Sugiura's [4] study, the data used here show the monthly number of prefecture-wise deaths due to the epidemic cold from 1918 to 1920, as stated in the JIDCS.[3] The term "epidemic cold" is used interchangeably with the flu in Japan. However, the number of deaths from the epidemic cold in the JIDCS is approximately 220,000 for those 3 years, a number that is significantly lower than the 385,000 deaths shown in "Ryukosei kanbo" ("Epidemic cold") mentioned above. This difference may be due to the complications between the flu and other diseases not included in the figures for the epidemic cold in the JIDCS. However, there is no reference to a definition of the epidemic cold in "Ryukosei kanbo," which may be considered relatively unreliable.

Meanwhile, Hayami [2] estimated the number of deaths from complications of the flu and other diseases based on data from the JIDCS and suggested adding them to the number of deaths directly from the flu, leading to a death toll from the Spanish flu exceeding 453,000. Although this method is highly appraised, it can only provide estimated figures, and uncertainties cannot be eliminated entirely. Consequently, the present study uses data from the JIDCS for analysis because the basic patterns of flu diffusion can be examined even if the number of deaths attributed to complications is excluded from the analysis.

12.2 Changes in the Number of Deaths Nationwide and Prefecture Mortality Rates from the Spanish Flu

Here, we consider the changes in the number of deaths nationwide due to the Spanish flu. For this study, the prefectures of Shizuoka, Nagano, and Niigata, and those east of them, are included in eastern Japan, while the remaining prefectures are included in western Japan.

12.2.1 Changes in the Number of Deaths Nationwide

Table 12.1 shows the number of deaths from the flu over 5 years, including 1918–1920 and the year before and after, and all deaths during the same period (hereafter, "all-cause deaths"). The table shows that in 1917, fewer than 100 people

---

[3] Only Table 12.1, described later, additionally uses data on the number of deaths in 1917 and 1921 (Bureau of Cabinet Statistics [11]; Bureau of Statistics [8–10, 12]). The compilation and publication of the JIDCS differ from year to year; however, this is due to the reorganization of the Bureau of Cabinet Statistics (now the Bureau of Statistics of the Ministry of Internal Affairs and Communications), essentially compiled by the same organization.

Table 12.1 Numbers of influenza and all-cause deaths, 1917–21

Year	Number of influenza deaths			Number of all-cause deaths		
	Male	Female	Both sexes	Male	Female	Both sexes
1917	36	32	68	609,310	590,359	1,199,669
1918	34,488	35,336	69,824	753,392	739,770	1,493,162
1919	21,415	20,571	41,986	648,984	632,981	1,281,965
1920	53,555	54,873	108,428	720,655	701,441	1,422,096
1921	5,229	5,075	10,304	659,328	629,242	1,288,570

Source: "Japan Imperial Death Cause Statistics"

died from the flu, including men and women; however, in 1918, the number increased by approximately 1000 times. After exceeding 100,000 in 1920, the number drastically dropped to one-tenth this level in 1921, confirming that the flu epidemic's explosion occurred from 1918 to 1920. The gender difference is minimal in all years, and considering these annual data, it is reasonable to assume that the flu spread did not involve gender selectivity.

In contrast, the number of all-cause deaths of both men and women during the same period was less than 1.2 million in 1917, exceeded 1.4 million in 1918 and 1920, and decreased back to 1.2 million in 1921. Despite combining the periods of the 1910s and 1920s to derive the average number of all-cause deaths (approximately 1.21 million) for the 20 years surrounding 1920, the main change can be understood as a value influenced by 1918 and 1920. Furthermore, the number of all-cause deaths and the mortality rate (27.3 and 25.4‰, respectively) in 1918 and 1920 were the highest and second highest, respectively, since 1899, when vital statistics began to be tabulated.[4] Subtracting the above average (approximately 1.21 million) from the number of all-cause deaths over the 3 years from 1918 to 1920, the total difference is approximately 570,000. Although this number significantly exceeds the number of flu deaths (approximately 220,000) based on the JICDS, it roughly reflects Hayami's [2] estimate (approximately 453,000). Therefore, considering demographic transition, the number of all-cause deaths in the 3 years from 1918 to 1920 is regarded as an "abnormal value" due to the special factor of the flu epidemic.

Fig. 12.1 shows two distinct waves in the flu epidemic during these 3 years under focus, hereafter referred to as the first and second waves. Both waves indicate that the flu was prevalent from winter to spring; however, the first wave peaks in November while the second wave peaks in January (2 months later). Regarding the pattern of change in the number of deaths, this decreased monotonically after the peak in the second wave; however, the first wave had a soft peak in February after its peak in November. The Spanish flu is now known to be caused by the H1N1 influenza A virus. However, it has been observed that the virus may have mutated in the second wave [7].

---

[4] Note that the demographic statistics lack data for the period 1944–46 when a large number of deaths occurred due to World War II.

Fig. 12.1 Monthly number of deaths due to Spanish flu, 1918–20. (Source: "Japan Imperial Death Cause Statistics"). (Note: The author calculated all the mortality rates from "Japan Imperial Death Cause Statistics.")

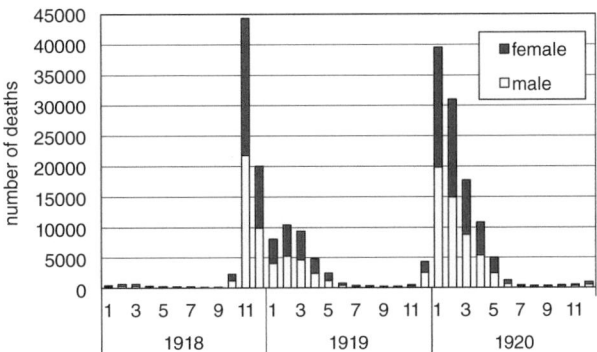

12.2.2 Changes in Mortality Rate by Prefecture

12.2.2.1 Calculation Method of Monthly Mortality Rate by Prefecture

When calculating monthly mortality rates by prefecture, the periods of the first and second waves were considered to be from September 1918 to August 1919 and from September 1919 to August 1920, respectively.[5] The rates were calculated by dividing the monthly deaths in each prefecture by the population of that prefecture at the ends of 1918 and 1919 for the first and second waves, respectively.

Additionally, it is recommended that statistics be standardized by age when discussing regional differences in mortality rates. However, for the Spanish flu, high deaths were reported among young and middle-aged people [2 pp. 76–77]; therefore, age standardization may not create the expected impact and was not included in this study.

12.2.2.2 Characteristics of the First Wave

The mortality rates (number of deaths per 100,000 people) were examined from the first wave at two-time points: the rate at the incipient stage (October 1918), when the number of deaths began to rise nationwide, and the rate at the peak stage, which varied from prefecture to prefecture.

Figures 12.2 and 12.3 show the distribution of mortality rates at the incipient and peak stages of the first wave, respectively, including the male and female mortality rates. At the incipient stage, prefectures with more than five deaths per 100,000 people were observed in western Japan for both males and females, and the Kinki region, Kagawa, and Tokushima had particularly high mortality rates. This aligns with Sugiura's [4] findings that the Spanish flu began to spread from major ports in

---

[5] This division was made because the number of deaths due to the Spanish flu was lowest in August or September in all of the years 1918, 19, and 1920.

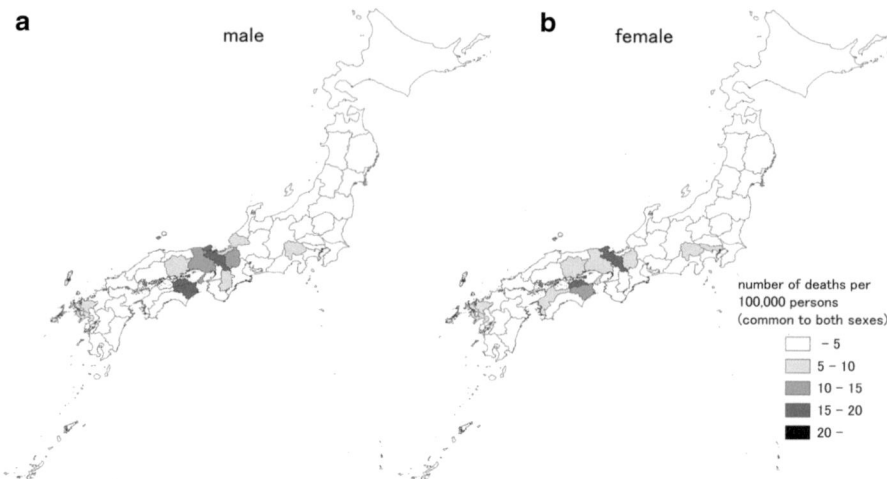

Fig. 12.2 Distribution of mortality rates at the incipient stage of the first wave. (**a**) male. (**b**) female. (Note: The author calculated all the mortality rates from "Japan Imperial Death Cause Statistics.")

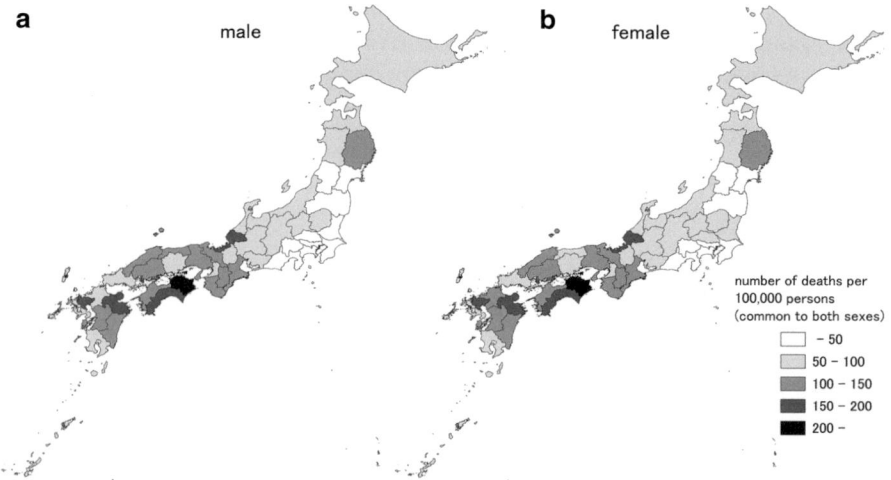

Fig. 12.3 Distribution of mortality rates at peak stages of the first wave. (**a**) male. (**b**) female. (Note: The author calculated all the mortality rates from "Japan Imperial Death Cause Statistics.")

western Japan. Additionally, the mortality rate in western Japan was generally high for males and females at the peak stage; however, the distribution pattern significantly differed from the incipient stage. This suggests that prefectures with slow epidemic transmission did not necessarily have lower peak mortality rates or that prefectures with early epidemic transmission did not necessarily have higher peak mortality rates. For example, in prefectures away from metropolitan areas, such as Iwate, Shimane, Kochi, Kumamoto, Oita, and Miyazaki, despite the expected low mortality at the incipient stage and slow transmission of the flu epidemic, there were relatively high peak mortality rates. Notably, Iwate is the only prefecture from eastern Japan among these.

12.2.2.3 Characteristics of the Second Wave

The mortality rates (number of deaths per 100,000 people) were also examined from the second wave at two-time points: the rate at the incipient stage (December 1919), when the number of deaths began to rise nationwide, and the rate at the peak stage, which varied by prefecture.

Figures 12.4 and 12.5 show the distribution of mortality rates at the incipient and peak stages of the second wave, respectively, including the male and female mortality rates, which were generally high in western Japan; however, some prefectures

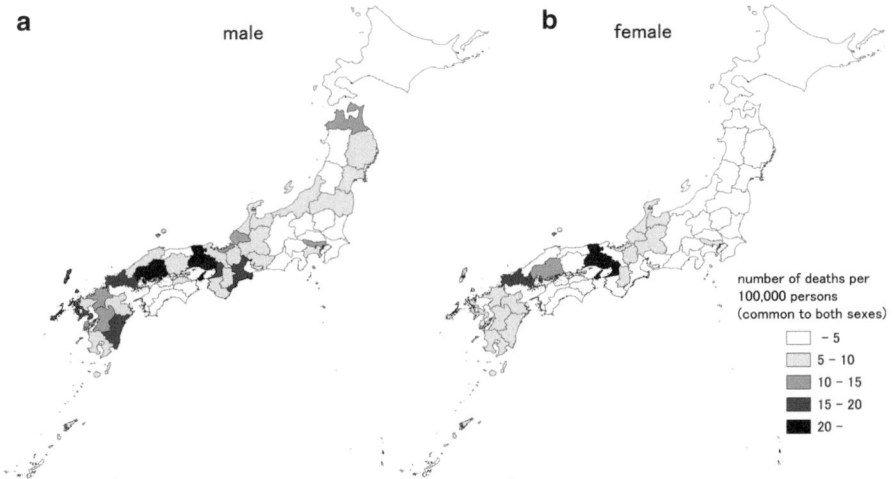

Fig. 12.4 Distribution of mortality rates at the incipient stage of the second wave. (**a**) male. (**b**) female. (Note: The author calculated all the mortality rates from "Japan Imperial Death Cause Statistics.")

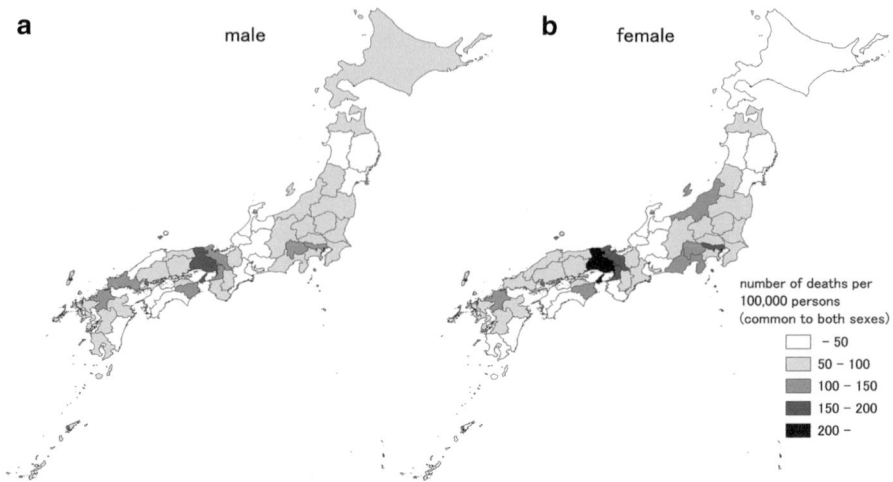

Fig. 12.5 Distribution of mortality rates at peak stages of the second wave. (**a**) male. (**b**) female. (Note: The author calculated all the mortality rates from "Japan Imperial Death Cause Statistics.")

showed high mortality rates at the incipient stage for males in eastern Japan, such as the Tohoku region and Niigata. Some prefectures, such as Mie, showed a significant gap between genders. Compared with the incipient mortality rate of the first wave, Kagawa and Tokushima showed low mortality rates, while Hiroshima, Yamaguchi, and Kyushu Island had high mortality rates. Regarding the peak mortality rate, no significant differences were observed between western and eastern Japan for either males or females. In other words, although the peak mortality rate was high in the Kinki region for males and females, it was also high in many densely populated areas, such as Tokyo, Yamanashi, Fukuoka, and surrounding areas. Among males, some prefectures, such as Mie and Miyazaki, showed relatively high mortality rates at the incipient stage; that is, they showed relatively low peak mortality rates despite the presumed early spread of the flu epidemic.

12.2.2.4 Comparison Between First and Second Waves

The months and mortality rates at the peak stages of the first and second waves were examined, and their characteristics were compared. No significant differences were observed between males and females in the distribution of the flu mortality rates, so only trends in males are discussed here.

Fig. 12.6 shows how many months each prefecture's peak month was delayed relative to the nationwide one for men, with (a) and (b) representing the first and second waves, respectively. In this figure, the numbers in the legend indicate the interval in months between each prefecture's peak month and the nationwide peak. Therefore, prefectures with zero values in the legend have the same peak month as

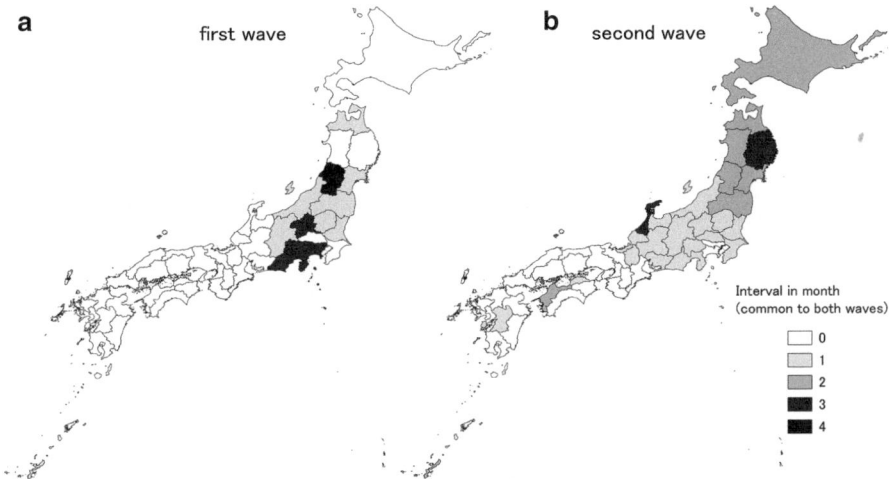

Fig. 12.6 Interval in months between each prefecture's peak month and the nationwide one (male). (**a**) first wave. (**b**) second wave. (Note: The author calculated all the mortality rates from "Japan Imperial Death Cause Statistics.")

the entire country. As shown in Fig. 12.1, the nationwide peak months are November 1918 for the first wave and January 1920 for the second wave.

Fig. 12.6 shows that in the first wave, all prefectures in western Japan had the same peak month as the entire country, obviously indicating that the flu spread earlier in western Japan. However, even in eastern Japan, the prefectures of Hokkaido, Iwate, Akita, Saitama, and Chiba had the same peak months as the whole of Japan, suggesting that the flu spread did not necessarily progress according to distance from western Japan.

The second wave's peak trend shown in Fig. 12.6 is similar to that of the first wave because the flu generally spread earlier in western Japan. Still, it differs from the earlier wave in that the peak months in Tokyo and Kanagawa coincided with those of the entire country. The virus is believed to have spread faster in western Japan and the Tokyo metropolitan area for the second wave. This aligns with Sugiura's [4] finding that Yokohama Port became a starting point for the flu spread. Additionally, in all prefectures in the Tohoku and Hokkaido regions of eastern Japan, away from the Tokyo metropolitan area, the peak month was 2 or 3 months later than in the whole country, differing from the first wave. Furthermore, in several prefectures in western Japan, the peak months were delayed by one or 2 months from that of the entire country, which was also different from the first wave.

Finally, based on Fig. 12.7, we consider the correlation between the peak mortality rates (deaths per 100,000 persons) of the first and second waves. Figure 12.7 plots Japan's 47 prefectures with the peak mortality rates of the first and second waves on the horizontal and vertical axes, respectively. As shown in the figure, the first-wave mortality rate exceeds that of the second-wave in the prefectures plotted in the lower right and vice versa in the prefectures plotted in the upper left. The

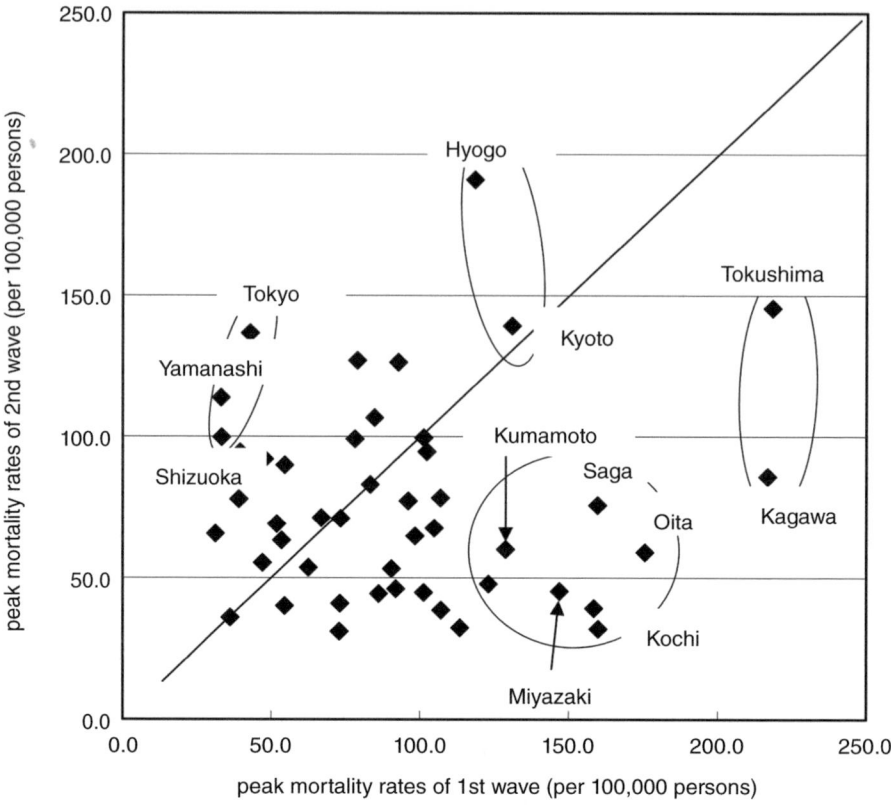

Fig. 12.7 Correlation between peak mortality rates of the first and second waves. (Note: The author calculated all the mortality rates from "Japan Imperial Death Cause Statistics.")

groups of prefectures in the actual vicinity of each other are also largely located closely together in the figure. Of these, the four groups located relatively far from the diagonal line are enclosed in a frame, with the prefecture's name also indicated. Of these four groups, the two located in the lower right include prefectures in western Japan: one comprising two prefectures in the Shikoku region and the other comprising seven prefectures centering on the Kyushu region, far from the three major metropolitan areas (Tokyo, Nagoya, and Osaka). In contrast, the two groups located in the upper right comprise prefectures in the three major metropolitan areas or in the periphery of these areas (i.e., the Kyoto and Hyogo group and the Tokyo, Yamanashi, and Shizuoka group). Accordingly, the peak mortality rate of the first wave exceeded that of the second wave in the prefectures belonging to the non-metropolitan areas of western Japan. Moreover, the opposite trend was observed in the three major metropolitan areas or prefectures adjacent to them.

12.3 Mathematical Model Application

Here, we classify the 47 prefectures into five types based on the monthly flu mortality rates, and by calculating the cumulative mortality rate for each type and applying the Gompertz and logistic curves to the cumulative mortality rate, we suggest mathematical models for describing the diffusion process of Spanish flu deaths. As shown above, no significant differences were observed between men and women in the flu mortality rate. Therefore, we calculated the mortality rate from the total number of deaths for men and women without giving any consideration to gender.

12.3.1 Method

Studies exploring the diffusion process of various infectious diseases, including the flu, through mathematical models, have primarily been conducted in the field of geography. For example, Sugiura [13] and Pyle [14] used the Monte Carlo method to explain the flu diffusion process. Furthermore, in the theoretical epidemiology field, which handles mathematical models of various mechanisms of infectious diseases, several studies have investigated the infectious disease diffusion process. These include, for example, studies on the flu by Spicer [15], Cliff et al. [16], and Nakaya [17] and studies discussing the spread of measles, Hodgson's disease, and Acquired Immunodeficiency Syndrome (AIDS).[6] According to Nakaya [18], this field has also been called "spatial epidemiology." These studies mostly used mortality or morbidity as variables to measure the extent of disease spread.

Meanwhile, growth curves such as the Gompertz and logistic curves are suitable for phenomena such as the spread of goods and fashion. The mechanism of such dissemination and transmission is strongly determined by the probability of contact between individuals. From a micro-scale perspective, evidently, this mechanism is common to the spread of infectious diseases. Therefore, we model it by applying these two curves to flu data. Cumulative mortality rates must be used to fit the flu diffusion process to these curves. However, to our knowledge, no study on the spread of infectious disease has used cumulative mortality or morbidity rates.

Both the first and second waves can be examined using the Gompertz and logistic curves. These two curves are suitable for describing phenomena strongly determined by the probability of contact between individuals, as described above. Typically, morbidity with infectious disease can be largely explained by contact between individuals; however, deaths from infectious diseases may be from various factors after contracting the infectious disease. Therefore, these two curves are inherently more suitable for morbidity than mortality. Moreover, according to Hayami [2], the morbidity rates of the Spanish flu were nearly 40% in the first wave and just over 4% in the second wave. The mortality rates among those affected were

---

[6] See Nakaya [14] for a discussion of these groups of studies.

12.1‰ in the first wave and 52.9‰ in the second wave. For the first wave in particular, there is a significant gap between morbidity rates and mortality rates among those affected. Therefore, the mortality rate of the first wave may include various factors other than contact between individuals, and the mortality rate of the second wave is more likely to reflect the morbidity rate. Furthermore, looking at the pattern of actual mortality rates, as mentioned above, the first wave showed a peak in November 1918, followed by a weaker peak in February 1919 (Fig. 12.1), and there is a high possibility that it deviates from the model. Hence, this chapter targets the second wave for modeling. Although this study assumes that the second wave of the flu epidemic started in September 1919 and ended in August 1920, these 2 months are excluded from the target months of modeling because the number of deaths is extremely few for those months. Hence, the period modeled is 10 months, from October 1919 to July 1920. In other words, ten observations are input per area.

The area unit to which the above two curves are applied is not by prefecture but by area group comprising multiple prefectures, that is, by type. This approach was taken because, in this modeling, the number of observations to be fitted to the curve is as small as 10, and if the area units were prefectures, the special circumstances of individual regions might have appeared. The classification of prefectures was performed using Ward's method of cluster analysis based on monthly mortality rates by prefecture. The number of clusters was determined so that each cluster's number of prefectures exceeded two, resulting in five regional groups. After calculating each group's average monthly mortality rate, the cumulative mortality rate was calculated for the target period of 10 months.

12.3.2 Classification of Prefectures Based on Mortality Rate

The clusters of prefectures are called Type 1 to Type 5 in descending order of the maximum value of the cumulative mortality rate, that is, the cumulative value in July 1920. The prefectures comprising each type are as follows:

Type 1: Five prefectures of Tokyo, Kyoto, Osaka, Hyogo, and Fukuoka.
Type 2: Six prefectures of Saitama, Chiba, Yamanashi, Shizuoka, Tokushima, and Kagawa.
Type 3: Fourteen prefectures of Hokkaido, Aomori, Yamagata, Fukushima, Ibaraki, Tochigi, Gunma, Niigata, Toyama, Nagano, Shimane, Ehime, Oita, and Okinawa.
Type 4: Ten prefectures of Kanagawa, Shiga, Nara, Tottori, Okayama, Hiroshima, Yamaguchi, Saga, Nagasaki, and Kumamoto.
Type 5: Twelve prefectures of Iwate, Miyagi, Akita, Ishikawa, Fukui, Gifu, Aichi, Mie, Wakayama, Kochi, Miyazaki, and Kagoshima.

Of these, Type 1 obviously includes metropolitan core areas and Type 2 includes the fringe districts of those core areas. In contrast, the remaining three types cannot be so easily classified. However, geographically, Type 3 is concentrated in the southern Tohoku, northern Kanto, and Shin'etsu regions, Type 4 is in the Chugoku and Kyushu regions, and Type 5 is in the Tohoku, Tokai, and Hokuriku regions. Subsection 12.3.4 discusses the characteristics of change in cumulative mortality for each type.

12.3.3 Application of Gompertz and Logistic Curves

The Gompertz and logistic curves are both types of growth curves primarily used to describe variables that increase in an S-shaped form. The Gompertz and logistic curves are expressed by Eqs. (12.1 and 12.2), respectively.

$$y = k \exp(-a \exp(-bx)) \quad (12.1)$$

$$y = \frac{k}{1 + a \exp(-bx)} \quad (12.2)$$

Here, k, a, and b are positive parameters.[7] These two equations are derived from the following differential equations.

$$\frac{dy}{dx} = Ay \exp(-bx) \quad (12.3)$$

$$\frac{dy}{dx} = By(k - y) \quad (12.4)$$

Here, A and B are both positive constants. As obvious from Eqs. (12.1 and 12.2), the y-values of the Gompertz and logistic curves converge to the parameter k if x approaches infinity. In other words, parameter k indicates the limit value of y. In any curve, parameter a determines the position in the horizontal axis direction, and parameter b represents the degree of curve slope, that is, the degree of increase in momentum. Additionally, the Gompertz and logistic curves have common features in that they increase monotonically and have a single inflection point (i.e., where the curve changes from "convex downward" to "convex upward"). However, while the logistic curve is symmetrical about the inflection point, the Gompertz curve does not have such symmetry.

Although used to explain processes such as the spread of goods and fashion, these curves have somewhat special effects on the flu diffusion process, which differs from typical processes of spreading. As Nakaya [17] observed, flu epidemics are generally more rapid from the outbreak to the peak than from the peak to the end. This may partly be due to the vast number of people who do not have immunity

---

[7] Although it is not absolutely necessary for all of these parameters to be positive for the Gompertz or logistic curve to hold true, the parameters are often treated as positive values when handling actual data.

to the flu during the early stages of a flu epidemic, referred to here as the initial effect. Additionally, it is easier to be infected with the flu virus in the dry winter season, referred to here as the winter effect. To fit data with such special effects to a growth curve representing the standard diffusion process, the influence of such effects must be removed from the data. Consequently, we propose the following two modified models.

First, to eliminate the initial effect's influence, the starting month is assumed to be September 1919, and the natural logarithm of the number of months that elapsed since September 1919 is assumed to suppress the epidemic speed from the outbreak to the peak. In this method, t is the number of months that have passed since the starting month, and the elapsed time is transformed into the natural logarithm of t ($\ln t$, $t > 0$) and then applied to the equation for the curve. By this variable transformation, the intervals of the observed values on the horizontal axis become wider during the early stages of the epidemic, suppressing the speed of the epidemic's spread. Here, the model with this modification is referred to as Modified Model I. A mathematical formula expressing this model can be obtained by substituting $x = \ln t$ ($t > 0$) into the Gompertz curve (Eq. 12.1) and the logistic curve (Eq. 12.2). Subsequently, the equations for the Gompertz and logistic curves are simplified to Eqs. (12.5 and 12.6), respectively.

$$y = k \exp\left(-at^{-b}\right) \tag{12.5}$$

$$y = \frac{k}{1 + at^{-b}} \tag{12.6}$$

Obviously, the y-values of these equations converge to the parameter k if t approaches infinity, as with Eqs. (12.1 and 12.2), and converge to zero if t approaches +0, unlike Eqs. (12.1 and 12.2).

Next, to eliminate the influence of the winter effect, the time difference between a given month and the coldest month is evaluated using the natural logarithm to suppress the speed of epidemic spread in winter. Letting t be the number of months from the starting month to a given month and s be the number of months from the starting month to the coldest month, then there is a method in which the elapsed time x is subjected to variable transformation, such as Eq. (12.7), and then applied to the equation for the curve.

$$x = \ln s(t - s + 1) \ \text{(when } t \geq s\text{)}, \quad x = \ln \frac{s}{s + 1 - t} \ \text{(when } t < s\text{)} \tag{12.7}$$

The starting and coldest months are assumed to be September 1919 and January 1920, respectively. Here, the model with these modifications is called Modified Model II. The mathematical notation of Modified Model II is given by substituting $s(t - s + 1)$ (when $t \geq s$) or $s/(s + 1 - t)$ (when $t < s$) in place of t in Eqs. (12.5 and 12.6).

12 The Diffusion Process of Spanish Influenza Deaths in Japan: Mathematical Models...

Table 12.2 Results of applying Modified Models I and II

Model name	Statistics	Type 1	Type 2	Type 3	Type 4	Type 5
Gompertz curve-based Modified Model I	k	2.87	2.78	2.17	2.07	1.34
	a	12.11	15.97	14.60	13.80	14.66
	b	2.17	2.18	2.00	2.21	2.20
	adjusted R^2	0.925	0.929	0.903	0.939	0.913
Logistic curve-based Modified Model I	k	2.87	2.78	2.17	2.07	1.34
	a	889.20	2168.51	942.44	800.81	685.18
	b	4.31	4.57	3.99	4.21	4.07
	adjusted R^2	0.861	0.914	0.979	0.956	*0.994*
Gompertz curve-based Modified Model II	k	2.87	2.78	2.17	2.07	1.34
	a	5.69	7.94	7.75	6.61	7.13
	b	1.31	1.35	1.24	1.35	1.35
	adjusted R^2	*0.955*	*0.995*	0.973	*0.997*	0.980
Logistic curve-based Modified Model II	k	2.87	2.78	2.17	2.07	1.34
	a	160.77	408.95	227.28	161.61	148.09
	b	2.50	2.73	2.39	2.48	2.40
	adjusted R^2	0.784	0.907	*0.993*	0.922	0.983

Note: All values were calculated by the author. Adjusted R^2 in italics shows the highest of four values in the same type.

12.3.4 Revised Model Application

Modified Models I and II based on the Gompertz and logistic curves were applied to the cumulative mortality values of Types 1–5 using the least squares method. Table 12.2 shows the application results, where k, a, and b mean parameters of Eqs. (12.1 and 12.2).[8] Additionally, the adjusted R^2 represents the goodness-of-fit with the highest value shown in italics.

According to Table 12.1, Modified Model II based on the Gompertz or logistic curve shows the highest fitness in the four types other than Type 5. This is because, for the Spanish flu, the curve fits better when the variables are changed to suppress the infectivity in the winter than when the infectivity is suppressed in the early stages of the epidemic. In other words, the winter effect is relatively stronger than the initial effect. Therefore, we further discuss Modified Model II, which considers the winter effect. Figure 12.8 is a superimposition of the cumulative mortality rate (‰) for each type after performing the variable transformation (Eq. 12.7) used in Modified Model II. As shown in this figure, even though Types 1 and 2 have similar

---

[8] The parameter k (marginal value of y) was estimated in advance by an appropriate method, and the remaining parameters a and b were calculated by the least square method; therefore, k is the same for all models, independent of the model form.

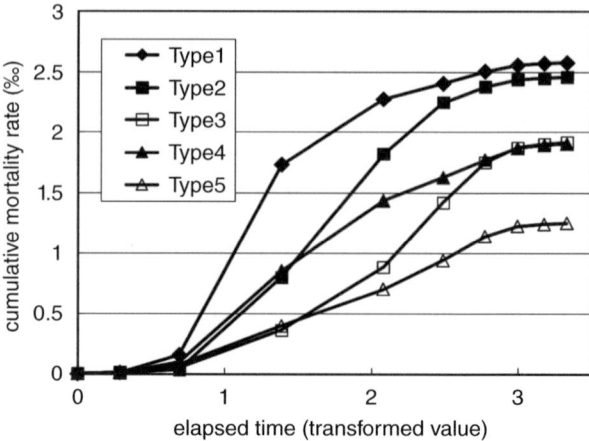

Fig. 12.8 Change in cumulative mortality rates for each type. (Note: The author calculated all the mortality rates from "Japan Imperial Death Cause Statistics.")

cumulative mortality rates at the end of the epidemic, there is a significant divergence in the values during the process. In other words, the cumulative mortality rate begins to increase earlier in Type 1 than in Type 2, indicating that the flu spreads faster in Type 1. A similar pattern is observed for Types 3 and 4, indicating that the flu spreads faster in Type 4. As previously mentioned, Type 1 is located in metropolitan core areas, whereas Type 2 is located in their fringes, and Type 4 is relatively closer to metropolitan areas than Type 3. This suggests that the second wave of Spanish flu first occurred in the metropolitan core areas of Tokyo, Osaka, and Fukuoka and then spread to the surrounding areas and further afield.

Figure 12.9 shows the regression curves drawn for Modified Model II and observed values for each type. Here, the regression curves of the Gompertz curve-based and logistic curve-based models are shown for Types 1, 2, and 4 and for Types 3 and 5, respectively. This is due to the selection of the curve with the better fit. As a result, the types with a relatively high limit value of y, that is, the parameter k, were combined with the Gompertz curve-based model, and the other types were combined with the logistic curve-based model. Regarding the relationship between the parameters given in Table 12.2 and the regression curve shape in Fig. 12.9, the limit value k evidently determines the vertical height of the curve. However, no clear relationship was observed between a, which determines the horizontal position, and the regression curve shape, or between b, which indicates the curve slope, and the shape. The degree of deviation of the observed values from the regression curve is minimal for Types 2–4, where the adjusted coefficient of determination exceeds 0.99. In contrast, for Type 1, which has a relatively low goodness-of-fit, the deviation is conspicuous at the fourth and fifth observed values from the left, and the Gompertz curve does not trace the sharp increase in cumulative mortality from December 1919 to January 1920. This is ascribed to an explosive flu epidemic in January 1920 in Type 1, comprising the metropolitan core areas of Tokyo, Kyoto, Osaka, Hyogo, and Fukuoka. A modified model that suppresses the winter effect more substantially is crucial to cope with such changes. Specifically, an evaluation method for the time difference between a given month and the coldest month using a function other than the natural logarithm might be considered.

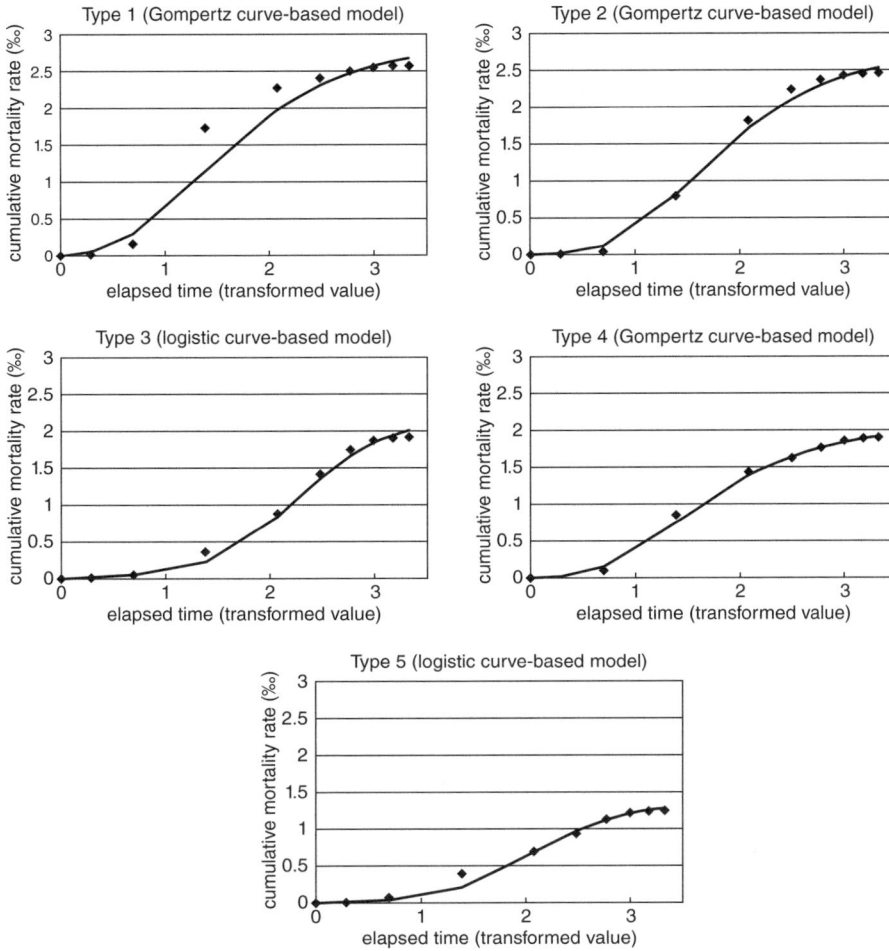

Fig. 12.9 Regression curves drawn using Modified Model II for each type. (Note: The author calculated all the mortality rates from "Japan Imperial Death Cause Statistics.")

12.4 Conclusion

This chapter examined the diffusion process of deaths caused by the Spanish flu pandemic using mortality statistics from Japan. For verification, Sect. 12.2 outlined changes in mortality rates by prefecture and Sect. 12.3 conducted a model analysis of the second wave of Spanish flu. Regarding model analysis, based on the Gompertz and logistic curves, this study assumed initial and winter effects during the flu epidemic and incorporated these two effects in Modified Models I and II, respectively. These can be proposed for each Gompertz or logistic curve, with four modified models prepared in total. In applying these models, prefectures were classified into five types; the average monthly mortality rates were calculated for each type, and the calculated cumulative mortality rates were then used as variables of those models.

Applying the above-modified models showed the following results. First, the fit of Modified Model II was typically better than that of Modified Model I, indicating that suppressing the momentum of the winter epidemic is more suitable in a model than the momentum of the early epidemic. Regarding the second wave of Spanish flu, the epidemic at its peak (i.e., January 1920) was explosive. Second, Modified Model II fits the Gompertz curve appropriately in areas with relatively high population density, indicating that the logistic curve tends to fit in areas away from them. The Gompertz curve begins to rise in value earlier than the logistic curve, suggesting that the second wave of Spanish flu first engulfed metropolitan areas simultaneously and then spread to the surrounding areas. Third, in Modified Model II, areas following the Gompertz curve typically have higher final cumulative mortality rates than those following the logistic curve. However, there are cases, such as Type 3 and Type 4, where the final cumulative mortality rates are similar but the upward curves differ. Hence, further verification is required.

As described above, among the two models presented, Modified Model II showed a good degree of fit. Therefore, when signs of a new flu epidemic appear in the future, the subsequent spread of the epidemic may be predicted using Modified Model II based on the Gompertz or logistic curve. Since the high influenza mortality rate among the elderly is a significant issue in contemporary Japan, the prediction using Modified Model II can be expected to contribute to improving that mortality rate and advancing gerontology. However, the following limitations must be considered to improve the accuracy of such predictions, that is, to improve the degree of model fit. First, we should consider transforming the elapsed time using a function other than the natural logarithm. Second, in this chapter, the start of the epidemic was subjectively determined, but because the start time can crucially impact the model, a method that objectively sets the time must be considered. Third, since the monthly mortality rate curve resembles the lognormal distribution, it is also significant to modify the model around it. Finally, given that the modified model presented in this chapter is an inductive model obtained by transforming variables to suit the cumulative mortality rate curve, to provide a distinct theoretical basis, a deductive method must be introduced, such as that used in theoretical epidemiology for modeling. In other words, it is vital to explain the macroscopic phenomenon of infectious disease epidemics through the microscopic phenomenon of viral transmission between individuals.

Acknowledgments This chapter is based on the following original paper published in Japanese: Inoue T (2010) Supein infuruenza ni yoru shibo no kakusan katei (The diffusion process of Spanish influenza deaths in Japan). In: Takahashi S., Nakagawa S. (eds) Chiiki jinko kara mita nihon no jinko ido (The demographic transition in Japan from the viewpoint of the regional population), Kokon Shoin, Tokyo, 77–98. I am grateful to Kokon Shoin for permitting me to translate and republish the paper.

References

1. Kaji M. Century of influenza: from "Spanish flu" to "bird flu". Tokyo: Heibonsha; 2005.
2. Hayami A. The Spanish influenza that hit Japan: the first world war between humans and viruses. Tokyo: Fujiwara Shoten; 2006.
3. Crosby AW. America's forgotten pandemic: the influenza of 1918. 2nd ed. Cambridge: Cambridge University Press; 2003.
4. Sugiura Y. Spatial diffusion of Spanish influenza in Japan, 1916-1926. Geogr Rev Japan. 1977;50:201–15. https://doi.org/10.4157/grj.50.201.
5. Central Sanitary Bureau. Department of Home Affairs (JP). Ryukosei kanbo (epidemic cold). Tokyo: Central Sanitary Bureau, Department of Home Affairs; 1922.
6. Okada H, editor. The threat of highly virulent new influenza. Fujiwara Shoten: Tokyo; 2006.
7. Ikeda K, Fujitani M, Nadaoka Y, Kamiya N, Hirokado M, Yanagawa Y. Precise analysis of the Spanish influenza in Japan. Ann Rep Tokyo Metr Inst P H. 2005;56:369–74.
8. Bureau of Statistics (JP). 1918 Japan imperial death cause statistics. Tokyo: Bureau of Statistics; 1921. https://dl.ndl.go.jp/en/pid/966023/1/1. https://dl.ndl.go.jp/en/pid/966024/1/2.
9. Bureau of Statistics (JP). 1919 Japan imperial death cause statistics. Tokyo: Bureau of Statistics; 1922. https://dl.ndl.go.jp/en/pid/966025/1/1.
10. Bureau of Statistics (JP). 1920 Japan imperial death cause statistics. Tokyo: Bureau of Statistics; 1923. https://dl.ndl.go.jp/en/pid/1939633/1/5.
11. Bureau of Cabinet Statistics (JP). 1917 Japan imperial death cause statistics. Tokyo: Bureau of Cabinet Statistics; 1920. https://dl.ndl.go.jp/en/pid/966021/1/1. https://dl.ndl.go.jp/en/pid/966022/1/1.
12. Bureau of Statistics (JP). 1921 Japan imperial death cause statistics. Tokyo: Bureau of Statistics; 1924. https://dl.ndl.go.jp/en/pid/966026/1/2.
13. Sugiura Y. Interurban diffusion of Asian influenza in Nagoya and its environs: a case study of spatial diffusion research. Geogr Rev Japan. 1975;48:847–67. https://doi.org/10.4157/grj.48.847.
14. Pyle GF. The diffusion of influenza: patterns and paradigms. Totowa: Rowman & Littlefield; 1986.
15. Spicer CC. The mathematical modelling of influenza epidemics. Br Med Bull. 1979;35:23–8. https://doi.org/10.1093/oxfordjournals.bmb.a071536.
16. Cliff AD, Haggett P, Ord JK. Spatial aspects of influenza epidemics. London: Pion; 1986.
17. Nakaya T. A spatio-temporal epidemic model for influenza: a case study of the 1988-1989 epidemic in Japan. Human Geogr. 1994;46:254–73. https://doi.org/10.4200/jjhg1948.46.254.
18. Nakaya T. Spatioepidemiological analysis. In: Nakaya T, Tanimura S, Nihei N, Horikoshi Y, editors. GIS for health care. Tokyo: Kokon Shoin; 2004. p. 74–126.

Part III
Preventive Medicine and Gerontology

Chapter 13
Measuring Facial Displacements and Strains for Cosmetics Development and Beauty Care

Satoru Yoneyama

Abstract Applications of a displacement and strain measurement method to facial skin for cosmetic development and beauty care are described. Displacement and strain are quantities that represent the movement of a point or the deformation of an object. Therefore, by measuring these quantities, deformation of the facial skin surface can be obtained, and the correlation between skin deformation and other factors, such as the skin moisture content and the amount of cosmetics, can be evaluated. Digital image correlation is used as a displacement and strain measurement technique. First, the results of strain measurements to investigate the relationship between wrinkles and strain that occur with aging are shown. Next, its application to the study of the effects of skin condition facial skin strain is described. Finally, an evaluation of the effect of facial massage by displacement measurement utilizing melanin obtained from ultraviolet light is presented. These results indicate that displacement and strain measurement can be used in cosmetics development and beauty care.

Keywords Strain · Displacement · Facial skin · Wrinkle · Massage · Cosmetics · Skin condition · Face lotion

---

S. Yoneyama (✉)
Department of Mechanical Engineering, College of Science and Engineering, Aoyama Gakuin University, Sagamihara, Kanagawa, Japan
e-mail: yoneyama@me.aoyama.ac.jp

© The Author(s), under exclusive license to Springer Nature Singapore Pte Ltd. 2024, corrected publication 2024
T. Shiozawa et al. (eds.), *Gerontology as an Interdisciplinary Science*, Current Topics in Environmental Health and Preventive Medicine,
https://doi.org/10.1007/978-981-97-2712-4_13

13.1 Introduction

Displacement and strain are quantities that express the movement of a point or the state of deformation of an object, and are mainly used in the fields of structural mechanics and solid mechanics related to the design of machines and structures, such as aircraft, automobiles, bridges, and their constituent members [1]. In these fields, stress is mainly used as an indicator of strength, but stress is an internal force per unit area and cannot be measured. On the other hand, displacement and strain are geometrical changes, such as the movement of an object and changes in dimensions, and therefore are measurable quantities [2].

Meanwhile, since human skin is constantly moving, it is always under strain. Wrinkles on the face increase with age, and it is not surprising that there is a correlation between the strain that occurs daily and the wrinkles that appear with age. In addition, facial skin is hydrated daily with a variety of lotions, and it is expected that the state of strain differs between dry and well-hydrated skin, as well as between skin that has been coated with a cosmetic product. Facial massage tightens the face, and displacement and strain can be considered effective in quantitatively evaluating the massage effect.

In this chapter, a displacement and strain measurement technique and some examples of the displacement and strain measurement results for cosmetics development and beauty care are described.

13.2 Strain, Strain Measurement and Digital Image Correlation

13.2.1 Displacement, Deformation and Strain

A definite distinction must be made among displacement, deformation and strain. Displacement is the vector of the movement of a point. Therefore, displacement does not necessarily mean deformation. Deformation can be considered to have occurred when the distance between two displaced points changes. Deformation is a change in the shape or dimensions of an object, and the degree of deformation is expressed by strain. Strain is the ratio of the change in dimensions to the reference length before deformation. It is a dimensionless value obtained by dividing the change in length by the original length. This is called normal strain. It is positive when the length is elongated and negative when it is contracted. Strain is a second-order tensor quantity. The deformation state can be described only by the normal strains, which are the tensor eigenvalues and are called the principal strains. In other words, any deformation can be expressed in terms of elongation and contraction. On the other hand, from a different point of view, a square object can be thought of as having strain such that it becomes a parallelogram. Such strain is called shear strain.

13.2.2 Displacement and Strain Measurement

Strain can be measured because it can be obtained from the dimensions and changes of an object. The most common strain measurement method is the electrical resistance strain gage method. This method is not suitable for skin strain measurement because the gage must be attached to the object to be measured. Skin is soft, so it is appropriate to use non-contact measurement techniques instead of contact measurement methods such as strain gauges. Optical and image measurement techniques can be used to measure displacement and strain on the surface of facial skin in a non-contact manner. There are several such methods, including the moiré and speckle methods, but in solid mechanics, digital image correlation has recently been widely used [3]. In digital image correlation, a random pattern on an object surface is used as an indicator, and the displacement is determined by detecting the movement of the pattern before and after deformation through image processing. If the surface to be measured is flat, displacement can be determined by measuring with a single camera. In the case of a three-dimensional surface, such as facial skin, the displacement of the three-dimensional surface is determined by measurement using the principle of stereo vision. Once the displacement distribution is obtained, the strain is calculated by differentiating the displacement distribution. To perform this measurement, the surface to be measured must have a random pattern. If there is no pattern on the surface, it is necessary to apply an artificial pattern. The details of this method are described in the textbook [3] as well as many other review articles [4–6]. In addition, one can find references [7, 8] on the application of this method not only in mechanical and civil engineering but also in bioengineering and medical fields.

13.3 Strain and Morphology on Facial Skin Surface

The surface morphology of the skin changes with age, resulting in wrinkles. Wrinkles are a major cause of skin problems because they give the appearance of aging, and improving wrinkles has been a cosmetic issue. Therefore, much research has been conducted on skin, especially in beauty care. For example, there have been studies on modeling skin structure and buckling analysis [9], measuring the moisture content, elasticity, and elongation of skin [10], research on the differences in wrinkle conditions when opening and closing the eyes [11], research on displacement distribution before and after skin moisturization [12], and research on measuring wrinkle conditions as they change with aging [13], and others. In general, wrinkles are caused by the effects of photoaging due to ultraviolet light, changes in skin structure with aging, and temporary wrinkles produced by daily facial movements. All of these influences are thought to act to change the skin surface morphology, although there are individual differences.

Wrinkles range from shallow and inconspicuous wrinkles to long and deep wrinkles. On the other hand, wrinkles do not tend to increase in number with age, as fine wrinkles tend to cluster with age to form deep, permanent wrinkles. Since these permanent wrinkles are difficult to improve with anti-wrinkle products, it is important to improve them before permanent wrinkles form.

Because skin characteristics vary among individuals, anti-wrinkle products do not always have the same effect; it is important to investigate which factors make a difference in the effectiveness of anti-wrinkle products. To conduct such an investigation, it is necessary to evaluate the changes in skin characteristics before and after using the product. In addition to using basic parameters such as water content and elasticity, it is also necessary to focus on the difference in skin movement with and without the effect of the product.

It can be considered that measurements focusing on skin movement will clarify the differences in skin movement with and without the effect of anti-wrinkle products and provide an approach for clarifying the mechanism of wrinkle formation. Here, the strain that is constantly generated by facial expressions is observed, and an example of strain measurement is shown.

13.3.1 Strain Caused by Daily Movements

To clarify the tendency of strain caused by daily activities, the results of strain measurements on four female subjects, one in her 20s, two in her 40s, and four in her 60s, are shown below. The measurement procedure can be found in Ref [14]. Figure 13.1 shows an image of the eye area of a subject in her 20s during blinking and an example of the strain distribution around the eye during this movement. A random pattern on the surface of the face is created by spraying black paint to apply digital image correlation, as shown in Fig. 13.1a. In Fig. 13.1b, the maximum principal strain, minimum principal strain, and principal direction distribution are shown. Unlike the common definition, the principal direction is the direction with the larger absolute value among the two principal strains. The principal strain distribution shows that the upper eyelid has tensile strain, and the lower eyelid has compressive strain. The direction of the crease line is orthogonal to the direction of the principal strain at the outer corner of the eye and parallel to the direction of the principal strain below the eye. The same tendency can be observed in the other subjects.

Although not shown in the figure, a comparison of the principal strain and the distribution of the principal direction among the four subjects reveals that although the strain values are different, the locations of strain generation and the distribution of the principal direction are generally consistent. The subjects in their 20s had more complex and varied principal directional distributions than the others, and almost no wrinkles were observed. Conversely, the principal directional distribution of other subjects is simple and varied smoothly, and wrinkles are observed. It is considered that the presence or absence of wrinkles can be related to the complexity of the principal strain direction distribution.

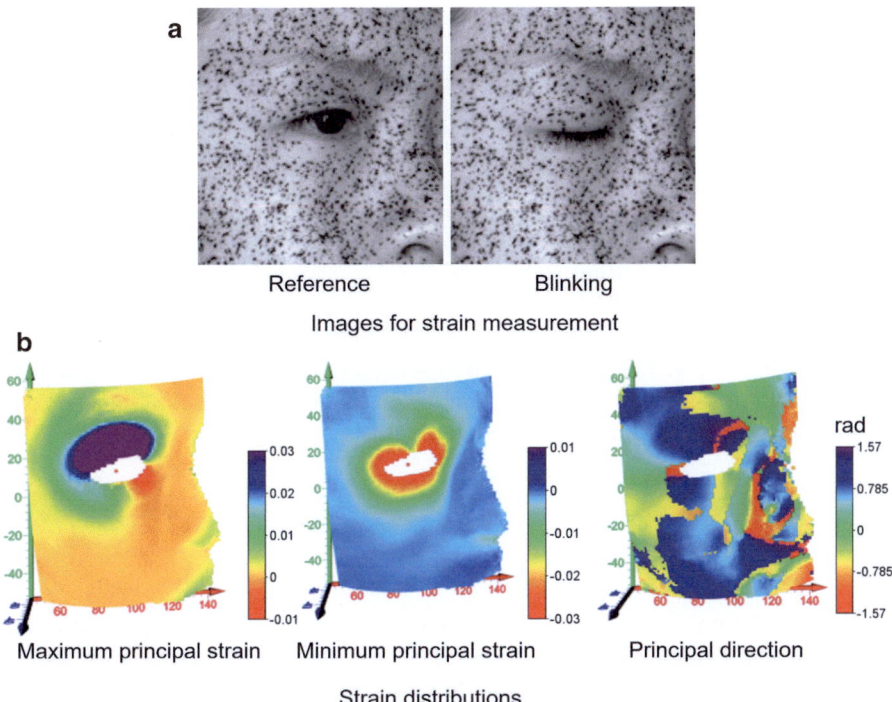

Fig. 13.1 Images for digital image correlation and strain distribution around the eye of a subject in her 20s

13.3.2 Effect of Moisturizing Condition on Strain Distribution

Next, the results of the investigation of the effect of skin conditions on strain distribution are described. Here, the strains on the facial skin surface during the blinking motion when the skin is normal, dry, and moisturized are investigated, and the changes in the strain distribution for different skin conditions are examined. The subjects are eight women in their 30s to 50s with an average age of 41.6 years.

Strains under the subjects' normal skin condition are defined as week −1 (−1 W). Next, strains under dry skin conditions without skin care for 1 week are designated as week 0 (0 W). The strains are examined after the application of serum. After 1 week of continuous use of the serum, the facial strain is the first week (1 W).

Although there are individual differences in the water content under and at the corners of the eyes, the water content tends to decrease the most when the skin is dry at 0 W and to increase toward 1 W. Comparing the distribution of the maximum principal strain generated when the eyes are closed in a blinking motion immediately after the application of the serum and when the eyes are dry at 0 W, it is found that the strain distribution tends to be more uniform after moisturizing than when dry. In other words, the strain becomes uniform throughout the entire surface after moisturizing. The strain distribution generated during blinking at 0 W and 1 W shows that the strain distribution at 0 W is more localized and concentrated, while the strain distribution at 1 W is more uniform immediately after moisturizing, as described earlier. Comparing the principal directional distribution with the lines of wrinkles on the 0 W skin surface, a relationship is observed in which the lines are orthogonal at the corners of the eyes and parallel under the eyes. The results of the present measurement test also show that the relationship between wrinkles and strain is different between the outer corner of the eyes and under the eyes since the locations of strain generation and the principal directional distribution tend to be almost the same in all subjects. Comparing the skin surface morphology between 0 W and 1 W, it is found that deep wrinkles with a clear direction on the skin surface are improved at 1 W. The subjects with improved skin surface morphology tend to have more uniform strain, suggesting that the change in strain distribution is related to the improvement in skin surface morphology.

13.3.3 Summary

The strains during daily activities are shown to clarify the relationship between the strain generated on the facial skin surface and the skin surface morphology. Strains are found to occur at the same location in all subjects. Strain values vary from person to person and are considered to be dependent on skin movement rather than skin characteristics. Around the eyes, the relationship between strain and skin surface morphology differs between the outer corner of the eyes and under the eyes. When the skin is dry, the strain tends to be localized, while when the skin is moisturized, the strain tends to be uniform over the entire area, and the skin surface morphology also tends to improve. As described above, when the skin is moisturized, the strain is generated uniformly over the entire surface, and the skin surface morphology may be improved. Please refer to references [15, 16] for details of the measurement results.

13.4 Skin Condition and Strains

Many women routinely use cosmetics from morning to night for long periods of time and feel changes in the physical properties of their skin and the state of their cosmetic coats, such as skin tightness and dryness. Therefore, there is a need for an objective method to measure the physical properties of cosmetic coats applied to facial skin. Many cosmetic coats are flexible from the viewpoint of usability and exist as thin coats on the skin surface. Therefore, contact measurement may not be able to acquire the condition of the applied skin, and non-contact measurement is desirable.

The skin is subjected to various physical stresses, such as dryness and other skin conditions. Thus, strain may change depending on the state of the cosmetic coating on the skin. Digital image correlation requires a characteristic pattern on the measurement surface. In the previous section, the strain is measured by applying a pattern by spray painting. Still, the pattern affects skin movement and strain may not be correctly acquired when assessing minute skin changes caused by cosmetic coats. In addition, if a pattern such as paint is applied after the cosmetic coat is applied, the cosmetic coat may be destroyed. Therefore, it is desirable to make measurements without using painted patterns.

Therefore, in this section, a technique to measure the local strain distribution near the eyes, which is deformed by blinking and its change, is developed. Strain near the eyes is focused on because many people perceive the skin around the eyes in several skin conditions. Digital image correlation is used as the strain measurement technique, and the measurement uses the microscopic morphology of the skin surface, such as micro relief, rather than a sprayed pattern. Since this study focuses on a localized area of the skin surface, the surface is assumed to be two-dimensional for evaluation. A method that can handle large deformations by analyzing a series of images is used to obtain the strain distribution on the skin surface and its changes. A method to obtain strain from images, including eyelashes, is also used. These methods are applied to several skin conditions and the measurement of strain near the eye during blinking in human subjects, demonstrating their validity. The results show that the strain measurement method can evaluate differences in skin conditions by measuring strain. It is also expected that the method can be used to evaluate skin conditions that change depending on cosmetics and other factors.

13.4.1 Measurement and Data Analysis

13.4.1.1 Setup for Measurement

In the experiment, three monochromatic high-speed cameras placed before the subject are synchronized to capture images near the left and right eyes and the entire face when blinking. Although images near the left and right eyes are recorded, the results for the right eye only are shown below. The camera recording local images near the eyes has a telephoto lens with a focal length of 200 mm, and the camera for the entire face has a zoom lens with a focal length of 28–300 mm. Images near the eyes are used to evaluate strain, while images of the entire face are used to extract images at the moment of blinking. The subject's face is mounted on a face stand, and images of the same area are captured in multiple skin conditions. The distance between each camera and the subject's face is approximately 450 mm, and the distance between the two cameras is approximately 600–1000 mm, depending on the shape of the subject's face. The subject's face is brightly illuminated by an LED flat light.

13.4.1.2 Measurement Procedure

The subjects are four healthy Japanese women of four consecutive generations, ranging from their 20s to 50s. After washing their face with the specified cleanser, the specified lotion is applied to the skin surface of the subjects in a prescribed amount. The subjects are then placed in a room where the temperature (20 °C) and humidity (40%) are kept constant. After 15 min of acclimatization, a high-speed camera is used to capture natural blinking behavior in a local area approximately 15 mm 13 mm below the eyes for 8 s at a frame rate of 200 frames/second. In addition, a prescribed foundation is applied to the subject's skin surface. After 15 min of acclimation, natural blink movements are captured in the same manner.

13.4.1.3 Digital Image Correlation Strain Analysis

Two-dimensional digital image correlation is used to measure the strain in the vicinity of the eye during blinking. From the 1600 images (200 frames/second × 8 s) for each subject in each condition, 101 images are extracted, including one blinking motion. An open-eye image is used as the reference image (before deformation). In digital image correlation strain analysis, the location of subsets in the deformed images can usually be found by drawing random patterns on the measurement area. In this study, micro-relief of the skin is used as the pattern for digital image correlation to avoid the painted pattern affecting the skin surface strain. Figure 13.2 shows the micro relief of four subjects. Micro relief can be seen on the skin surface of all subjects. The area for evaluating strain distribution is approximately 8.7 mm × 6.5 mm.

Because the skin is easily stretched and deformed, displacement cannot be determined if the blinking motion is large or if the skin surface is hidden by long eyelashes. To solve these problems, two analysis algorithms are developed: the "reference update algorithm" used for larger deformations and the "eyelash removal algorithm."

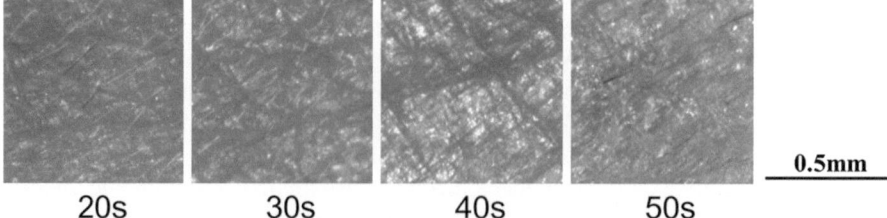

Fig. 13.2 Skin micro-relief of four subjects

13.4.1.4 Reference Image Update Algorithm

When the number of detected points in the current image is less than 90% of the total data points, the reference image is replaced with the next image and a new reference image, called the updated reference image, is used. This updating process is repeated until the number of detected points becomes 90% or more. The displacement of the current image is then obtained by adding the displacement between the updated reference image and the original reference image and the displacement between the current image and the updated reference image. By performing this operation automatically, the displacements of the images can be obtained continuously, even if the deformation is large.

13.4.1.5 Eyelash Removal Algorithm

The skin surface of the analysis area may be obscured by eyelashes during the analysis, and the same gray level distribution as the subset in the reference image may not be detected. In this measurement, the gray levels in the skin images ranged from about 80–200. Therefore, in the data processing of digital image correlation, pixels in the subset with a gray level of less than 60 are not used to calculate the correlation coefficient, and the remaining gray values are used to find the position of the subset. In this case, if the entire subset is filled with eyelashes, the position of the subset cannot be determined. Therefore, it is necessary to specify the size of the subset leaving the skin part so that the entire subset is not filled with eyelashes.

13.4.2 Strain Distribution Measurement

At first, the reproducibility of blinking motion and strain distribution is examined. The principal strain distribution under the right eye at the moment of eye closure before foundation application shows that strain measurement is possible through digital image correlation using the minute irregularities of the skin rather than a painted pattern. It is also observed that the same principal strain distribution is obtained for the same subject and skin condition after three repetitions of the blinking motion. This result indicates that this measurement's principal strain distribution of blinking motion is highly reproducible.

Figure 13.3 shows the distribution of the maximum and minimum principal strains and the principal direction of the right eye at eye-closed timing for a subject. The principal direction is the angle between the horizontal axis and the principal axis, which indicates the maximum principal strain. The maximum principal strain is almost positive, and the minimum principal strain is almost negative in the skin conditions before and after foundation application. The maximum tensile and compressive strains at the measurement site occurred close to the right eye. The principal direction results for both skin conditions before and after foundation application

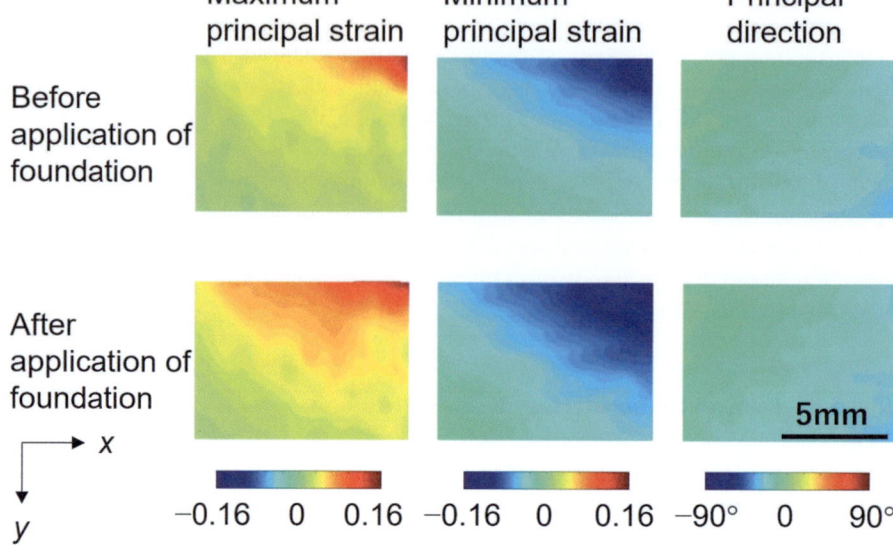

Fig. 13.3 Distributions of principal strains and principal direction before and after the application of foundation

show that the principal direction is tilted approximately −45 degrees (upper right) in all subjects, indicating that the skin surface is deformed toward the center of the eye.

It is not shown in this figure, but the results of the principal strain for four subjects show that the tensile deformation behavior differs from subject to subject. In one subject, the tensile area of the skin surface under the right eye extends from the eye in a fan shape, while in the other subjects, the tensile area extends horizontally. On the other hand, the results for the minimum principal strain show no significant differences in compressive behavior between subjects.

The distribution of the maximum principal strain is observed to change after the foundation is applied. However, the minimum principal strain and the principal strain direction do not change significantly after foundation application in all subjects. These results suggest that evaluating the maximum principal strain distribution is useful for visualizing differences in skin behavior in various subjects and skin conditions.

13.4.3 Time History of Strain

The principal strain time histories of the skin surface under the right eye during a single blink for a subject are shown in Fig. 13.4. The results show that the strain value increases for both subjects as the eyes are closed, reaching a maximum value

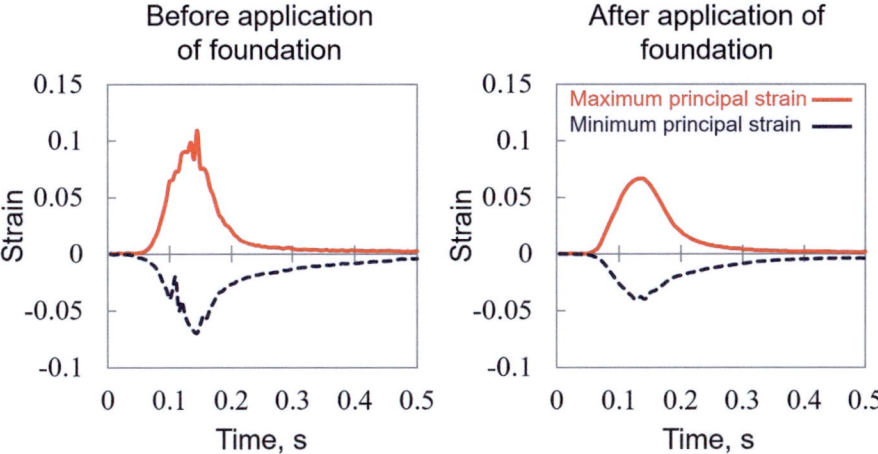

Fig. 13.4 Time-histories of principal strains before and after the application of foundation

at approximately 0.5 s and returning to 0 when the eyes are opened. The results for the subject show that the different values of the maximum and minimum principal strains are observed before and after the foundation. Therefore, the evaluation of the time history of the maximum and minimum principal strains is useful for visualizing the continuous change in strain values during blinking and the maximum absolute value of strain at the timing of eye closure. This evaluation makes it possible to compare the differences in skin behavior before and after the subject's foundation application.

13.4.4 Summary

In this section, a technique for measuring the local strain distribution near the eyes during blinking and its changes is examined. The micro-relief of the skin is used for measurement. A method that can handle large deformations by analyzing a series of images is developed to obtain the strain distribution and its change on the skin surface. In addition, a method to obtain the strain when eyelashes are included in the image is established. The effectiveness of the measurement method is shown by measuring strains near the eye during blinking for several skin conditions and subjects. The results show that the measurement method can evaluate differences in skin conditions. In the future, the strain measurement method is expected to be used to evaluate skin conditions that change depending on cosmetics and other factors. More details of this measurement technique can be found in Ref [17].

13.5 Evaluation of Massage Effect

Wrinkles and sagging skin that increase with age are major cosmetic problems and account for many women's skin problems. A person's appearance accounts for a large percentage of their impression, and the "face" is a very important element in determining that impression. The main causes of wrinkles and sagging are dryness and poor blood circulation. To solve these problems, oil massage has recently been used to improve blood circulation and reduce wrinkles and sagging by draining accumulated waste (lymph fluid). This is because improving blood circulation allows nutrients flowing in the blood to reach skin cells more efficiently and activates turnover (the cycle of skin cells from creation to shedding), thereby making wrinkles thinner. However, currently, the evaluation of such massage effects is mainly done by visually comparing two images before and after the massage, and a quantitative evaluation of the massage effect has not been done.

It has been shown that skin strain can be measured by digital image correlation. However, an applied pattern disappears because cream is used when an oil massage is performed. Therefore, conventional digital image correlation techniques cannot be applied to evaluate the effect of facial massage. On the other hand, it is known that existing melanin pigment under the epidermis is observed under the illumination of an ultraviolet light. It does not disappear due to the application of cream to the surface of the skin during the massage. Therefore, it can be used instead of an applied pattern to the skin's surface. The following is an example of evaluating the massage effect by measuring human skin displacement before and after massage.

13.5.1 Measurement of Massage Effects

In order to use the stereo method, two cameras are mounted vertically so that the face could be photographed by two cameras. The light sources are an ultraviolet LED light with a wavelength of 350 nm and a white LED light. The subjects are women in their 40 s. First, the subjects are asked to wash their face and then to take a temperature of 22 °C after washing their face. After washing their face, the subjects entered a constant-temperature, constant-humidity room with a temperature of 22 °C and a humidity of 50%. After changing the skin's moisture content in the constant-temperature, constant-humidity room, a photograph of the face before the massage is taken. After the massage, the face is photographed to complete the massage effect evaluation.

Figure 13.5 shows an example of an image taken of a subject and the displacement vectors. The eyes are hidden to protect privacy. From these images, it can be seen that melanin pigment appears due to the use of ultraviolet light as a light source. On the other hand, no changes in facial contours before and after the massage can be visually observed. A vector diagram of the three-dimensional surface

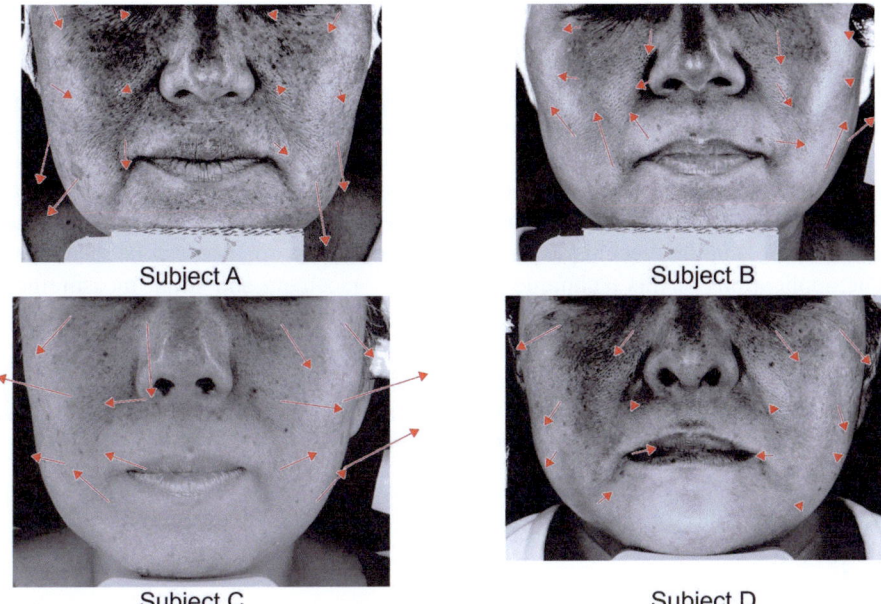

Fig. 13.5 Vector representations showing the effect of massage

displacement distribution obtained from these images using the stereo image correlation method shows that the facial skin is stretched in the left-right direction and contracted in the vertical direction by massage. These results indicate that facial massage causes changes in the position of the skin, such as the mouth to rise and the cheeks to move outward.

Compression is found to occur mostly in the face line around the chin and around the eyes. These areas are almost identical to the areas where major lymph nodes and lymphatic vessels are located. Therefore, although there are individual differences, it can be considered that the lymphatic fluid stagnating in the face is drained by the oil massage, and the swelling of the face is reduced, resulting in many compression distortions in the face line around the chin and in some parts of the cheeks.

13.5.2 Summary

The effect of massage can be evaluated by obtaining displacement distributions before and after massage through image correlation using ultraviolet images. If a method to quantitatively evaluate the effect of massage is introduced to the beauty industry in the future, it will lead to efficient training of estheticians, customer satisfaction, and increase people's quality of life. More details of this measurement results can be found in Ref [18].

13.6 Conclusions

This chapter shows three displacement and strain measurement applications of digital image correlation to cosmetics development and beauty care. Displacement and strain are quantities that are usually used in the field of structural mechanics and solid mechanics. They represent the movement of a point or the deformation of an object. The digital image correlation technique is used for measuring displacement and strain. Using this technique, facial skin surface deformation can be visualized. The applications of this technique to facial displacement and strain measurement show that the strain measurement technique is effective in cosmetics development and beauty care.

Acknowledgments The works described in Sects. 13.3 and 13.4 were supported by Kao Corporation and Sect. 13.5 by C'bon Cosmetics Co., Ltd.

References

1. Hibbeler RC. Mechanics of materials. 3rd ed. Upper Saddle River: Prentice-Hall; 1997.
2. Dally JW, Riley F. Experimental stress analysis. 3rd ed. New York: McGraw-Hill; 1992.
3. Sutton MA, Orteu J-J, Schreier HW. Image correlation for shape, motion and deformation measurements. New York: Springer; 2009.
4. Hild F, Roux S. Digital image correlation: from displacement measurement to identification of elastic properties – a review. Strain. 2006;42:69–80.
5. Pan B. Recent progress in digital image correlation. Exp Mech. 2010;51:1223–35.
6. Yoneyama S. Basic principle of digital image correlation for in-plane displacement and strain measurement. Adv Compos Mater. 2016;25:105–23.
7. Palanca M, Tozzi G, Crisofolin L. The use of digital image correlation in the biomechanical area: a review. Int Biomech. 2016;3:1–21.
8. Ferraiuoli P, Fixsen LS, Kappler B, Lopata RGP, Fenner JW, Narracott AJ. Measurement of in vitro cardiac deformation by means of 3D digital image correlation and ultrasound 2D speckle-tracking echocardiography. Mecical Eng Phys. 2019;74:146–52.
9. Kuwazuru O, Saothong J, Yoshikawa N. Mechanical approach to aging and wrinkling of human facial skin based on the multistage buckling theory. Med Eng Phys. 2008;30:516–22.
10. Krueger N, Luebberding S, Oltmer M, Streker M, Kerscher M. Age-related changes in skin mechanical properties: a quantitative evaluation of 120 female subjects. Skin Res Technol. 2011;17:141–8.
11. Tsukahara K, Hotta M, Osanai O, Fujimura T, Kitahara T, Takema Y. The effect of eye opening and closing on the result of facial wrinkle assessment. Skin Res Technol. 2009;15:384–91.
12. Staloff IA, Rafailovitch M. Measurement of skin stretch using digital image speckle correlation. Skin Res Technol. 2008;14:298–303.
13. Bazin R, Lévêque JL. Longitudinal study of skin aging: from microrelief to wrinkles. Skin Res Technol. 2011;17:135–40.
14. Miura N, Sakamoto T, Aoyagi Y, Yoneyama S. Visualizing surface strain distribution of facial skin using stereovision. Theor Appl Mech Lett. 2016;6:167–70.
15. Miura N, Arikawa S, Yoneyama S, Koike M, Murakami M, Tanno O. Digital image correlation strain analysis for the study of wrinkle formation on facial skin. J Solid Mech Mater Eng. 2012;6:545–54.

16. Miura N, Murakami M, Koike M, Arikawa S, Yoneyama S, Tanno O. Strain measurement on facial skin surface using 3D digital image correlation (strain and morphology on facial skin surface). Trans JSME. 2013;79:774–8.
17. Sakai K, Zhang Y, Yoneyama S, Miyazaki Y, Hanada Y, Nagai Y, Igarashi T. Evaluating distribution and variation of strains near eyes in blinking using digital image correlation. Skin Res Technol. 2020;26:749–59.
18. Zhang Y, Sunamura F, Bamba S, Arikawa S, Yoneyama S, Motonami K, Matsushima T, Numata T. Evaluation and visualization of facial massage effects by using ultraviolet stereo-image correlation. Skin Res Technol. 2020;26:349–55.

Chapter 14
Educational and Collaborative Model for Early Detection and Intervention of Age-Related Hearing Loss to Enhance Health and Well-Being of the Aged

Tomoko Sano, Noriko Katsuya, Hisao Osada, and Keiko Morita

Abstract Age-related hearing loss is one of the most common chronic conditions; however, many people do not immediately see a doctor when they notice hearing loss. Untreated hearing loss can affect older adults' health and quality of life. In this chapter, we first describe age-related hearing loss, the physical and psychosocial effects, and how to cope with hearing loss, including the use of hearing aids and communication strategies. Following this, we will discuss the difficulties of early detection and intervention from the viewpoint of the Japanese social system and a standard of care that is not well known. Thirdly, we discuss the results of our survey conducted on local government officials throughout Japan, which revealed that few local governments in Japan conduct hearing tests for the elderly and that some local government officials do not know enough about hearing loss or how to deal with people with hearing difficulties. Based on the above, we next propose an educational and collaborative model that enables the early detection of hearing loss and intervention to help those with hearing difficulties. In this proposed model, municipalities, researchers, and physicians will work together to educate the general public, other municipalities, and schools to spread knowledge about hearing loss and hearing aids. In addition, local government officials and elderly members of local

---

T. Sano (✉)
Josai International University, Chiba, Japan
e-mail: sano@jiu.ac.jp

N. Katsuya
The University of Tokyo, Tokyo, Japan

H. Osada
J. F. Oberlin University, Tokyo, Japan

K. Morita
Daito Bunka University, Saitama, Japan

senior citizens' clubs who have received special education, will take the role of spreading knowledge to the next step. Finally, we introduce the Finger Rub/Finger Tap screening test for the early detection of hearing loss. This is a test that anyone can perform quickly and efficiently.

Keywords Age-related hearing loss · Screening test · Healthcare · Community · Hearing education · Collaboration

14.1 Age-Related Hearing Loss and Its Impacts

According to the World Health Organization's (WHO) critical facts on deafness and hearing loss, over 1.5 billion people globally live with hearing loss, which has the potential to increase to over 2.5 billion people by 2050 [1]. Furthermore, the prevalence of hearing loss increases with age; among individuals older than 60 years, approximately 30% have hearing loss [1]. In turn, age-related hearing loss can significantly affect older adults [1]. It is estimated that, in Japan, more than 15 million people aged 65 years and older have mild hearing loss or more [2]. The proportion of people with hearing loss increases with age, with 84.3% of men and 73.3% of women aged 80 or older presenting with this condition [2]. The current aging rate in Japan is 29.1% [3], and is expected to reach 33.3% in 2036 and 38.4% in 2065 [4]. As the population ages, the number of older people with hearing loss will continue to increase and the health issues related to this condition will become increasingly severe.

14.1.1 What Is Age-Related Hearing Loss?

Age-related hearing loss (presbycusis) is defined as age-dependent progressive, bilateral, and symmetrical sensorineural hearing loss, which first occurs for sounds of high frequencies [5]. As this condition gradually progresses, it is difficult for patients to notice the hearing loss [6]. Age-related hearing loss is a complex disorder caused by both genetic and environmental factors [6, 7]. It results from the cumulative damage sustained by the patient over a lifetime due to multiple factors [5, 8]. It possibly results from the shedding or degeneration of hair cells within the cochlea in the inner ear; sclerosis of blood vessels, which play an important role in maintaining the cochlear function; decreased elasticity of the basal plate; and changes in the central nervous system [6, 9]. Additionally, diet, exercise, years of exposure to noise, and the type of lifestyle of an individual play a role in the loss of hearing. The sensorineural tissues of the cochlea have very limited repair capacity [8]. Therefore, once the cochlear receptor cells, hair cells, and cochlear neurons have been damaged, they do not regenerate, making cellular loss permanent [8]. Sensorineural hearing loss is considered incurable. However, hearing loss can be prevented to some extent by maintaining good health via a well-balanced diet, moderate exercise, and good sleep as well as avoiding noise exposure, such as listening to loud music for long periods [5, 6].

Age-related impairments of auditory information processing, such as sound source localization, temporal resolution, and word-tone listening ability, have been previously observed [6]. Owing to the difficulty in hearing high-frequency sounds, it becomes harder to discriminate consonants with aging, resulting in a high incidence of mishearing words [6]. It has often been observed that older patients with hearing loss can hear words but cannot recognize them. Hearing words becomes particularly problematic when the speaker speaks quickly or in the presence of background noise, such as the noise of a crowd of people [10]. In addition, the recruitment phenomenon also occurs, in which quieter sounds are less audible and louder sounds are heard significantly louder, i.e., the range of the perfect sound level becomes narrower.

Hearing loss, which significantly affects the life and health of individuals, is associated with increased healthcare costs and economic losses. The WHO states that neglected hearing loss costs the world $750 billion annually and that interventions such as preventing, identifying, and treating hearing loss are cost-effective and provide significant benefits to individuals [1]. Therefore, hearing loss is a global health challenge that needs to be addressed by all nations. This is especially urgent in Japan, where the population is aging rapidly.

14.1.2 Relationship Between Age-Related Hearing Loss and Health Problems

Many previous studies have reported an association between hearing loss in old age and other health problems.

14.1.2.1 Hearing Loss and Other Physical Concerns

According to a large study of older adults aged 65 years and older performed at the University of Michigan, hearing loss is associated with a risk of systemic disease [11]. Older people with hearing loss have an increased risk of developing arthritis (1.41 times), cancer (1.35 times), cardiovascular disease (1.48 times), emphysema (1.41 times), hypertension (1.29 times), stroke (1.39 times), and other diseases compared with those without hearing loss [11]. However, the mechanisms underlying these associations are not understood. Whether these health problems can be improved by enhancing hearing health care is also a subject for future investigation.

Several studies have reported an association between sleep apnea and hearing loss [12, 13]. For example, in a Japanese case, CPAP therapy in a 50-year-old man with sleep apnea and hearing loss improved his apnea as well as his hearing ability, which improved from an average hearing loss of 38.8 dB to one of 13.8 dB [13].

Dalton et al. (2003) performed a large survey of 2688 individuals aged 53 years or older to examine the relationship between hearing loss and quality of life (QOL) in old age [14]. Pure tone audiometry was performed to assess the hearing level. Additionally, activities of daily living (ADL), instrumental activities of daily living (I-ADL), and SF-36 were used as health-related quality-of-life indicators, and the Hearing Handicap Inventory for Elderly-S (HHIE-S, [15]) was used to measure difficulty in communication [14]. The results showed that severe hearing loss was significantly correlated with hearing handicaps and SF-36 psychological and physical functional decline [14]. In addition, individuals with moderate-to-severe hearing loss exhibited greater difficulty with ADLs and I-ADLs than those with normal hearing. Similar results were reported in the large study conducted by Gopinath et al. (2012) [16], in which individuals with moderate-to-severe hearing loss had 2.9 times more difficulties than individuals with normal hearing [16]. Hyams et al. (2018) found that individuals with impaired hearing who did not wear hearing aids had significantly lower SF-36 overall health than those with normal hearing [17].

Furthermore, a significant relationship was found between hearing loss and falls [18, 19]. For example, Lin et al. (2012) performed a cross-sectional study of the relationship between hearing loss and falls among participants aged 40–69 years as part of the National Health and Nutritional Examination Survey (NHANES) [19]. The participants were asked, "Have you had difficulty with falling during the past 12 months?" and were subjected to an objective vestibular balance test. A logistic regression analysis showed that, for every 10 dB of increase in the level of hearing loss, the likelihood of falling increased by 1.4-fold [19]. It is considered that the cognitive load caused by hearing loss, which reduces attention resources, resulted in the reported falls. Because that study was based on self-reports regarding falls and was a cross-sectional study, further investigation is needed.

Recent studies have investigated cross-modal plasticity in the brain [20–22]. These studies reported that when hearing loss occurs, the volume of the auditory cortex is reduced, and it is used for other information processing, such as visual information processing and somatosensory perception [20–22]. Cross-modal reorganization has been found to occur in the early stage of hearing loss [20, 22], which can be restored by administering early intervention, such as using adequately adjusted hearing aids [21].

14.1.2.2 Hearing Loss and Psychosocial Concerns

Psychosocial impacts of age-related hearing loss are well documented, including depression [23–25], social isolation [26], and social and emotional loneliness [27]. Saito et al. (2010) conducted a longitudinal study of 580 people aged ≥65 years without depressive symptoms and revealed that hearing handicaps could predict the development of depressive symptoms 3 years later [28]. Jayakody et al. (2022) investigated the association among untreated hearing loss, social and emotional loneliness, social isolation, social support, and psychological discomfort (depression, anxiety, and stress) in older adults using a cross-sectional study design [27]. Their findings revealed that untreated hearing loss significantly contributed to both moderate and intense levels of emotional loneliness but was not associated with social loneliness [27]. They concluded that hearing health professionals, such as audiologists, should be aware of the psychosocial burden that may accompany hearing loss [27].

Moreover, hearing loss affects cognitive function. Several studies have reported an association between hearing loss and dementia [e.g., 29, 30, 31]. Uhlmann et al. (1986) examined whether hearing loss could predict cognitive decline in patients with Alzheimer's disease. They observed that the decline in cognitive function after 1 year was approximately twice as pronounced in the hearing loss group than it was in the healthy-hearing group, with a statistically significant difference [29]. Subsequently, Lin et al. (2011) revealed that hearing loss is an independent risk factor for dementia; the risk of developing dementia was three times higher for those with moderate hearing loss and 4.94 times higher for those with severe hearing loss [30]. Thus, preventing hearing loss could be an approach to suppressing the development of dementia [31, 32]. The Lancet International Committee has identified hearing loss in middle age as the greatest part (8%) of the 40% modifiable risk of dementia [31].

There are four hypotheses on possible mechanisms that can lead to age-related hearing loss and cognitive decline: cognitive load on perception, sensory deprivation, cascade, and common cause hypotheses [32–34]. Future development in this area is strongly encouraged because aggressive treatment of hearing loss may contribute to delaying the onset of dementia [32, 33].

14.1.3 Dealing with Hearing Loss

14.1.3.1 Hearing Assistance

Early hearing replacement for hearing loss is necessary to prevent the negative physical, psychological, and social effects of hearing loss. The use of a properly adjusted hearing aid can improve the hearing ability of patients. Although the level of clarity in patients will not be restored to the level in their youth, they will be able to hear and understand words. Hearing aids can contribute to the resolution of speech communication problems. Hearing aids are also expected to have emotional (reducing anxiety, depression, and anger) [35] and cognitive (preventing or slowing the progress of dementia) benefits [32, 35]. Boi et al. (2012) studied the effects of hearing aids on mood, QOL, and the sense of burden of care among caregivers of elderly patients with hearing loss and depression, and they observed significant improvements in depressive symptoms and QOL [36]. The improvement occurred within a period as short as 1 month and lasted up to 6 months [36]. Furthermore, the burden of care decreased among caregivers.

Cochlear implants are used in cases of severe hearing loss that cannot be adequately compensated using hearing aids. Cochlear implants are expected to improve hearing loss and prevent emotional and cognitive declines in a similar manner as hearing aids.

14.1.3.2 Communication and Other Help

Other considerations, such as environmental and communication considerations, can help reduce stress and maintain good relationships. People with hearing loss have difficulty hearing in crowded or noisy environments; thus, it is crucial to reduce the ambient noise [37–39]. Devices that assist hearing aids, speech–text–transcribe software, and writing boards can also help in communication.

Many patients with hearing loss unconsciously lip-read, but rarely they use other adaptive strategies [39, 40]. Instead, they tend to use maladaptive strategies, such as pretending to understand or avoiding conversation. Moreover, problem-solving strategies considered adaptive, such as "ask until you know" and "seek expert advice," were not significantly related to mental health [40]. Therefore, research on communication strategies for people with hearing loss is warranted. Additionally, communication strategies should be developed for people around older people with hearing loss because communication is a reciprocal exchange.

We must remember that older patients will still face some hearing difficulties despite using hearing aids [37]. Several studies have mentioned the following communication cautions that should be practiced while communicating with older patients with hearing loss. As for behavioral aspects, "approach the older person," "get their attention before you speak," "tell them the topic before you speak," "turn toward them," and "speak in a quiet place and in a position where light illuminates the speaker's face" [37, 38]. Regarding speech aspects, "use simple words," and when they do not understand, "rephrase instead of repeat," or "write down," "speak clearly," "speak slowly," and "do not shout" [37–39]. These points for communication were mentioned for nurses to keep in mind while interacting with older patients with hearing loss [37, 38]. In addition to nurses, other healthcare professionals, family members, colleagues, friends, and acquaintances should keep these points in mind.

14.1.3.3 Relation Between Hearing Loss and Psychological Aspects

The association between hearing loss and poor mental health conditions, such as depression and anxiety, has been repeatedly reported in previous articles [23–25]. However, the mechanism underlying this relationship remains unclear. The use of a combination of available interventions, such as hearing aids and psychotherapy, to treat patients is desirable. Although clinical psychological interventions are necessary for treating older patients with hearing loss, effective psychotherapies have not been adequately tested. Hearing loss is a communication disorder, and traditional talk therapies are greatly affected by its nature. Because traditional psychotherapy is conducted through dialogue, people with hearing loss may miss or mishear what is being said. The possibility of improvements in the psychological state and hearing ability of the patient while using hearing aids should also be examined. Moreover, adaptive changes in self-stigma and perception of hearing aids through psychotherapy may increase the motivation in patients to wear and continue using hearing aids.

Psychological health may also be improved by providing psychoeducation to adapt communication strategies. Nevertheless, people around elderly patients with hearing loss should be supportive while communicating. A review by Cosh et al. revealed limited but positive evidence that audiological rehabilitation, including hearing aids and community-based hearing interventions, can improve mental health [25].

The method to provide clinical psychological treatment to elderly individuals with hearing loss is unclear. It is necessary that healthcare professionals should focus on assessing both hearing loss and depression [41]. Blazer (2018) reported that primary care physicians may be best placed to identify both hearing loss and depression in the first instance [41]. Only few studies have discussed the methods to approach emotional and mental health in the audiology setting [42]. Bennett et al. (2020) conducted a study on this issue. They investigated the current knowledge, beliefs, and practices adopted by Australian audiologists to address the emotional and mental health needs of adults with hearing loss [42]. They revealed that two-thirds of the audiologists were underconfident and felt they lacked the skills to provide emotional support to individuals with hearing loss (64.5%). The obstacles in the delivery of emotional support included "worry that I may get out of my depth (56.6%)," "time/caseload pressures (55.3%)," and the perception that providing emotional support was "not within the scope of practice of audiologists (31.6%)" [42].

Conversely, there is a need to consider the measures that can be adopted by clinical psychologists and other healthcare professionals to recognize hearing loss, refer to it when it is identified, and respond to the individuals with the condition. Mental health professionals must be more sensitive to patients with "silent" disabilities, which complicate the treatment of psychiatric disorders [41]. The therapeutic benefits of psychological treatment for hearing loss patients are also significant [41]. However, guidelines to screen asymptomatic older adults for hearing loss have not yet been established [41].

14.2 Challenges in Assisting Individuals with Age-Related Hearing Loss

14.2.1 Difficulty of Initial Intervention

The lack of understanding of hearing loss is a common problem in many countries, and the importance of early detection of this condition and early hearing replacement has been repeatedly pointed out. It has been reported that more than 10 years elapse between the time at which elderly persons first experience hearing loss and the time at which they consult with a hearing care professional [43, 44]. Because age-related hearing loss progresses gradually, many individuals are unaware of their condition. Instead, they tend to underestimate their hearing loss and perceive that their difficulty hearing is caused by the poor speaking style of the speaker [45].

Therefore, a help-seeking behavior is unlikely to occur [44]. Because old individuals tend to delay attending medical examinations, a screening system is needed in this setting.

David et al. (2017) performed semi-structured interviews of older adults with age-related hearing loss and reported the formation of the help-seeking process and attributional model [46]. The self-stigma associated with age-related hearing loss has three stages: awareness, agreement, and self-concurrence, which include three dimensions each: cognitive attribution, emotional response, and behavioral response [46]. Furthermore, those authors found that hearing aids significantly positively affected the experience of stigma in all stages and dimensions of self-stigma.

The reasons reported by older adults for not using hearing aids include their ineffectiveness, difficulty in adjustment, maintenance issues (battery replacement), stigma, and cost [2]. Morita et al. (2022) reported that AEldraSagen (which means "elder matter"), a national, large, volunteer, nonprofit organization, conducts various volunteer activities in Denmark. This organization has a hearing aid volunteer group in the city of Fredensborg. This group changes batteries and clean products [47]. Moreover, they work to change the negative perception of hearing aids and provide a hands-on experience of using them to help seniors realize their benefits and that good hearing improves their quality of life [47]. These activities can effectively reduce stigma and promote the continuous use of hearing aids, but few empirical studies have been conducted in this regard.

14.2.2 Challenges in Addressing Age-Related Hearing Loss in Japan

Difficulties in detecting age-related hearing loss and early intervention are also experienced in Japan. It is evident that the condition in Japan is lagging compared to that in other countries.

14.2.2.1 Current Situation of Age-Related Hearing Loss in Japan

According to Japan Track, the hearing aid ownership rate among the population with impaired hearing in Japan is approximately 15%, which has slightly increased since 2018 (14.4%); however, the ownership rate is still low [48]. The hearing aid ownership rates are over 40% in other countries; for instance, the rates in Denmark, the United Kingdom, Norway, and France are 55%, 53%, 49%, and 46%, respectively [48]. Approximately 51% of respondents in Japan received a hearing test within 5 years, primarily from their family doctor (28%) or otorhinolaryngologist (27%). However, 33% of respondents had never had a hearing test, which indicates that detecting hearing loss in its early stages is difficult due to the lack of a system mandating regular screening tests.

Of all hearing aid owners, 50% were satisfied with their hearing aids, and 51% wished they had used them earlier because they thought they would have had a more comfortable social life [48]. Conversely, those who did not own hearing aids had several reasons for not using them, including "annoyance," "the hearing loss was not so severe," "hearing aids would not restore the hearing," and "cannot afford to purchase hearing aids" [48]. Only 8% non-owners of hearing aids in Japan knew about the national hearing aid program that provides public subsidies for purchasing hearing aids [48]. Additionally, 75% of Japanese were unaware of cochlear implants. Among patients with hearing loss, only 14% knew a "certified hearing aid technician" [48]. These indicators show that many people in Japan have limited knowledge about hearing loss, hearing aids, cochlear implants, and support systems. The limitation includes the lack of awareness regarding the routes for purchasing hearing aids, hearing aid adjustments, and qualifications for hearing aid sales.

14.2.2.2 Lack of Standard Pathway of Care for Hearing Loss

Owing to some reasons, older Japanese patients with hearing loss do not seek medical attention, which leads to a low rate of hearing aid use. Hearing tests are performed during annual health checkups for employed people, unlike retired seniors who do not have the opportunity to undergo a hearing test.

The first problem in caring for hearing loss is the lack of social implementation of screening in Japan. Second, there is a lack of awareness regarding the information on places to go for consultation upon noticing hearing loss and the kind of treatment and support available when people have hearing loss. In Japan, there is a certification called speech-language-hearing therapist, which is a combination of an audiologist and a speech pathologist. Of Japan's 38,200 certified speech-language-hearing therapists, only 2097 specialize in hearing [49]. Moreover, approximately 5000 otolaryngologists (less than approximately half of otolaryngologists) are registered as hearing-aid consultation doctors who diagnose hearing loss, select a hearing aid, and determine hearing aid adjustments in this country [50]. There are currently few hearing specialists. In Japan, there are two methods to purchase hearing aids: (1) visit an otorhinolaryngology clinic and get the hearing aids from the clinic, and (2) visit a hearing aid store and meet a certified hearing aid technician in person. However, even in the latter case, the patient must first visit an otorhinolaryngologist to receive a "medical information form on hearing aid compatibility" and bring the form to the hearing aid store. Therefore, it is mandatory to first visit an otolaryngologist and get medically examined before purchasing a hearing aid. However, in reality, people can purchase hearing aids by mail order. In such cases, the hearing aids are not properly calibrated according to the patient's requirements, which is unknown to the common people. Most people do not know that the fitting process can take 3–6 months. It is also unknown to the public that a hospital or hearing-aid specialty store with a hearing-aid consultant or certified hearing-aid technician must be selected. It often happens that a hearing care professional is not an ear specialist or that there is no certified hearing-aid technician at the hearing-aid store; therefore, the hearing aid is purchased but not adjusted. As a result, many elderly people put ineffective hearing aids in their wardrobes because they have not been adjusted. These hearing aids are called wardrobe hearing aids.

14.2.3 Study: The Medical Checkup System and Knowledge of Hearing Loss among Workers in Japanese Municipalities

Purpose
As mentioned earlier, a screening system has not yet been implemented in Japan. Additionally, a nationwide survey has not been conducted in this regard. Therefore, this study aimed to investigate the extent to which and how screening for hearing loss in the elderly is conducted in Japan. If the extent of screening was found to be low, this study also aimed to clarify the reasons for the same. Furthermore, we examined the knowledge and perception of age-related hearing loss among local government officials. We investigated whether their daily communication with the elderly differed depending on their level of knowledge.

Methods
Subjects: Workers in charge of welfare for the elderly in municipalities in Japan.
Questionnaire: A questionnaire was developed that included the following items: Whether and why screening tests for hearing loss are offered to older adults, knowledge and perception about age-related hearing loss, and communication efforts with older adults with hearing loss. Whether or not hearing tests are conducted for the elderly; if yes, what is the examination fee, and what percentage of eligible persons were examined; if not, the reasons provided were (1) unknown, (2) it is not an immediate necessity, (3) unable to budget, (4) the tests were conducted earlier, but are no more conducted due to low number of examinees, (5) free answers, and further plans.

Five knowledge-based questions are listed as follows: (1) hearing loss should be taken care of as soon as possible, (2) hearing loss in older adults does not need to be addressed (invert scale), (3) I know where to buy hearing aids, (4) hearing aids need to be adjusted, and (5) hearing loss is a preventable risk factor for dementia. Eleven perception- and attitude-based questions are mentioned as follows: (1) age-related hearing loss is not a big deal and it just causes difficulty in hearing, (2) age-related hearing loss is not a big problem for older adults, (3) age-related hearing loss is a problem for the families of patients, (4) talking to an older person with hearing loss requires energy, (5) older adults are reluctant to admit their hearing loss, (6) hearing improves quickly using hearing aids, (7) I want to use hearing aids as soon as I develop hearing loss, (8) I want to know more about age-related hearing loss, (9) hearing loss is a personal matter that needs to be addressed by individuals, (10) education about hearing loss prevention is important, and (11) public authorities are not involved in hearing loss in the old residents. These questions were prepared to examine the knowledge and perceptions of municipal officials regarding age-related hearing loss. All surveys were conducted using a 5-point scale ranging from "very much agree" to "not at all agree. To investigate how the municipalities' staff communicate with the older people with hearing loss, the two-test (yes, no) method was used to obtain the answers to the 15 questions that were used based on the previous studies [37, 38, 40]. These questions enquired about the methods used to deal with elderly persons with hearing loss and are listed as follows: speak slowly, speak clearly, write down when they don't understand, keep sentences short, look at the person while you are speaking to, rephrase when they didn't understand, occasionally check to see if the person understands what you said, use plain words, use gestures, remove your mask to allow lip-reading, do not shout, move to a quieter environment, get their attention before you speak, let them know when you change the speaker, tell when the topic changes, and other open-ended questions.

Procedures: A request letter, questionnaire, and return envelope were sent to 1916 municipalities, and responses were obtained from 634 municipalities, without names (response rate: 33.1%).

Survey period: September–October 2019.

Ethical considerations: After approval by the Research Ethics Committee of Josai International University (approval number: 06 T190019), the purpose of the study was presented, and responses were obtained only when consent for research cooperation was obtained.

Results

All statistical processing was performed in IBM SPSS statistics version 23. Descriptive sttistics were applied to reveal the implementation of screening tests for hearing loss in the municipalities and the knowledge and attitudes of the staff regarding hearing loss and hearing aids.

Screening System

Simple tabulations revealed that only 14 local governments (2.1%) carried out hearing screenings for the elderly (Table 14.1). Of these, 12 (1.8%) offered the examinations free of charge at the same time as regular health checkups. Two organizations

Table 14.1 Status of hearing screening in municipalities

	Implemented n(%)	Not implemented n (%)	No answer n(%)	Total n (%)
Hearing screening is being conducted	**14 (2.1)**	**630 (96.3)**	**10 (1.5)**	**654 (100)**
Separately from regular checkups	2 (0.3)			
Simultaneously with regular checkups	12 (1.8)			
Free	11 (1.7)			
Paid	1 (0.2)			
No answer	2 (0.3)			
There are plans to implement in the future		2 (0.3)		
No plans to implement it in the future		615 (94.0)		
No answer		13 (2.0)		

Table 14.2 Categorized free descriptions of other reasons for not conducting hearing health examinations

Category	n	%
Lack of evidence (no legal basis, not in the medical checkup items, no evidence)	36	32
No awareness of need (never thought about it, no request from residents, no sense of need)	27	24
Other supports (individually handled, conducted at physical examinations)	20	18
Handled by different departments (implemented by the department in charge of disability)	14	13
Difficult to implement (difficult to secure budget and implementation system)	13	12
Other (consideration at the counter is necessary)	1	0.9

(0.3%) performed the hearing checkups separately from the regular checkups, and these organizations charged for the hearing checkups but did not indicate their cost. Only two organizations indicated the examination rate (6.8% and 100%). Both cities conducted the examinations simultaneously with the medical checkups, and the examinations were free of cost. The former was a city with a population of approximately 24,000, while the latter was a small village of 370. Only two organizations that had not yet implemented the system were planning to do so starting in 2020.

Regarding the reasons for not implementing the screening program, the respondents chose one of the following options: (1) unknown reason (419 [66.5%]), (2) not an immediate necessity (79 [12.5%]), (3) out of budget (52 [8.3%]), (4) previously implement the program but stopped due to a low number of examinees (1 [0.2%]), and other (109 [17.3%]).

The free descriptions of other reasons (109 responses, 111 sentences) were categorized (Table 14.2). The most common reason was "Lack of evidence (36 [32%])," followed by "No awareness of need (27 [24%])," and "other supports (20 [18%])."

Knowledge, Awareness, and Communication Skills
The mean and standard deviation of answers obtained for each knowledge- and awareness/attitude-based question was calculated (Table 14.3). The total scores for the five knowledge items were calculated, and GP analysis was conducted to divide the respondents into two groups: high and low knowledge groups. Since the respondents who answered "disagree" to the question, "hearing loss in older adults does not need to be addressed," were considered more knowledgeable than those who responded "agree," the scores for it were inverted. Respondents with total knowledge scores of ≤ 21 and ≥ 25 were added to the low and high knowledge groups, respectively.

A t-test was conducted with high and low knowledge as the independent variables and the perception and attitude items as the dependent variables to examine the effect of difference of knowledge on perception (Table 14.3). No differences were found in the knowledge of hearing loss with respect to the three items: talking with an older adult with hearing loss requires energy (low group; 3.73, high group; 3.89), public authorities are not currently involved in hearing loss in the old residents (low group; 3.28, high group; 3.27), and older adults are reluctant to admit

Table 14.3 Knowledge, perception, and attitude toward age-related hearing loss

Items	All (n = 654) Mean	All SD	Low (n = 204) Mean	Low SD		High (n = 188) Mean	High SD	t	p	95%CI
Knowledge of age-related hearing loss										
Hearing loss should be taken care of as soon as possible	3.82	0.78								
Hearing loss in older adults does not need to be addressed[a]	4.24	0.74								
I know where to buy hearing aids	3.74	1.02								
Hearing aids need to be adjusted	4.43	0.68								
Hearing loss is a preventable risk factor for dementia	4.14	0.77								
Perceptions and attitudes toward age-related hearing loss										
Age-related hearing loss is not a big deal, it just causes difficulty in hearing	2.13	0.86	2.54	0.844	>	1.64	0.713	11.375	***	0.741, 1.050
Age-related hearing loss is not a big problem for older adults	1.45	0.77	1.77	0.854	>	1.26	0.723	6.450	***	0.358, 0.671
Age-related hearing loss is a problem for the families of patients	4.19	1.06	3.91	1.058	<	4.47	0.961	−5.503	***	−0.762, −0.361
Talking to an older person with hearing loss requires energy	3.83	0.87	3.73	0.871		3.89	0.936	−1.783		−0.343, −0.17
Older adults are reluctant to admit their hearing loss	2.89	0.90	2.90	0.827		2.83	1.020	0.714		−0.118, −0.253
Hearing improves quickly using hearing aids	2.49	0.86	2.76	0.840	>	2.19	0.844	6.677	***	0.401, −0.736
I want to use hearing aids as soon as I develop hearing loss	3.14	0.82	3.22	0.780	<	3.62	0.867	−4.727	***	−0.560, −0.231
I want to know more about age-related hearing loss	3.70	0.71	3.34	0.708	<	3.97	0.682	−9.050	***	−0.773, −0.497
Hearing loss is a personal matter that need to be addressed by individuals	2.52	0.87	2.71	0.752	>	2.28	0.930	5.014	***	0.262, 0.600
Education about hearing loss prevention is important	3.72	0.74	3.44	0.689	<	4.07	0.674	−9.197	***	−0.769, −0.498
Public authorities are not currently involved in hearing loss in the old residents	3.28	0.88	3.28	0.811		3.27	1.042	0.137		−0.174, 0.200

Note: CI = confidence interval
*** $p < 0.001$
[a]The item was scored as a reversal item, with 1 being very much agreed upon and 5 not agreed upon at all

their hearing loss (low group; 2.90, high group; 2.83). Scores were relatively high for the responses obtained for items "use energy for conversation" and "public institutions are not involved." Significant differences were found for all other items. Overall, the scores for the following four items were relatively low, with the high knowledge group scoring significantly lower than the low knowledge group; "age-related hearing loss is not a big deal, it just causes difficulty in hearing (M(SD) 1.64(0.71), 2.54(0.84), $t(387)$=11.37, $p < 0.001$)," "age-related hearing loss is not a big problem for older adults (M(SD) 1.26(0.72), 1.77(0.85), $t(387)$=6.45, $p < 0.001$)," "hearing improves quickly using hearing aids (M(SD) 2.19(0.84), 2.76(0.84), $t(387)$=6.68, $p < 0.001$)," "hearing loss is a personal matter that need to be addressed by individuals (M(SD) 2.28(0.93), 2.71(0.75), $t(360)$=5.01, $p < 0.001$)." Those with more knowledge did not perceive hearing loss as a condition associated only with difficulty in hearing or not being a big problem for the elderly. Further, they did not perceive hearing loss as a problem that should be dealt only by the patients For the remaining four items, the high knowledge group scored significantly higher than the low knowledge group; "age-related hearing loss is a problem for the families of patients (M(SD) 4.47(0.96), 3.91(1.06), $t(390)$=−5.50, $p < 0.001$)," "I want to use hearing aids as soon as I develop hearing loss (M(SD) 3.62(0.87), 3.22(0.78), $t(387)$=-4.73, $p < 0.001$)," "I want to know more about age-related hearing loss (M(SD) 3.97(0.68), 3.34(0.71), $t(390)$=−9.05, $p < 0.001$)," "Education about hearing loss prevention is important (M(SD) 4.07(0.67), 3.44(0.69), t(389)=−9.20, $p < 0.001$)." Those with more knowledge strongly recognized that age-related hearing loss is a problem for their families, they want to use hearing aids if they develop hearing loss, and they want to learn more about age-related hearing loss. The more knowledgeable they are, the more they will understand the importance of hearing care. Furthermore, they believed in the importance of education to prevent hearing loss. Thus, education is necessary to understand hearing loss correctly and to respond appropriately, and a certain level of knowledge is necessary to understand and learn more about hearing loss.

For items related to communication, we calculated frequencies and percentages to determine the total and individual knowledge groups. A χ-square test was conducted to determine the presence of differences between groups (Table 14.4). In terms of overall responses, "speak slowly (90.2%)," "speak clearly (88.0%)," and "write down when they didn't understand (74.7%)" were the most frequently practiced communication methods. Conversely, those that were rarely practiced were "tell when the topic changes (9.8%)," "tell when the speaker changes (11.7%)," "getting their attention before starting to speak (17.5%)," "move to a quieter place (18.8%)," and "do not shout (19.3%)." The high-knowledge group was significantly more creative in their communication on all items than the other group ($p < 0.01$).

Discussion

The survey results revealed that very few Japanese municipalities (2.1%) were conducting hearing tests for older people at the time of the study. Moreover, the reasons for not doing so were "lack of awareness," such as "no urgency," "no legal basis," and "never having thought about it." Additionally, the awareness and attitude of local government officials toward age-related hearing loss varied depending on the

Table 14.4 Municipal officials' response to the old residents with hearing loss

Items	All n = 654		Low knowledge n = 204		High knowledge n = 188		P
	n	%	n	%	n	%	
Speak slowly	532	90.2	150	73.5	170	90.4	<0.001
Speak clearly	519	88.0	143	70.1	167	88.8	<0.001
Write down when they didn't understand	441	74.7	110	53.9	153	81.4	<0.001
Keep sentences short	378	64.1	88	43.1	137	72.9	<0.001
Look at the person while you are speaking	378	64.1	88	43.1	136	72.3	<0.001
Rephrase when they didn't understand	346	58.6	83	40.7	116	61.7	<0.001
Sometimes, check to see if the person understands what you said	341	57.8	77	37.7	131	69.7	<0.001
Use plain words	337	57.1	84	41.2	122	64.9	<0.001
Use gestures	255	43.2	51	25.0	104	55.3	<0.001
Remove your mask to allow lip reading	244	41.4	52	25.5	98	52.1	<0.001
Do not shout	114	19.3	20	9.8	55	29.3	0.001
Move to a quieter environment	111	18.8	14	6.9	58	30.9	<0.001
Get their attention before you speak	103	17.5	10	4.9	57	30.3	<0.001
Let them know when you change the speaker	69	11.7	8	3.9	37	19.7	0.001
Tell them when the topic changes	58	9.8	5	2.5	30	16.0	<0.01

amount of knowledge they had about hearing loss. Furthermore, high-knowledge staff were significantly more likely to respond appropriately when communicating with older adults with hearing loss. These findings suggest that correct knowledge about hearing loss must be widely disseminated to legalize hearing screening and provide appropriate responses to older people.

It is necessary to establish a law and a budget as soon as possible, provide health education and training to the public, and disseminate correct knowledge of the relationship between hearing loss and other health problems. The difficulty of early screening, the low rate of hearing-aid use, and the low level of satisfaction with hearing aids in Japan are attributable to the Japanese system of hearing impairment and the lack of knowledge about hearing loss among the public and government officials.

After this study was launched, the COVID-19 pandemic broke out, and we were forced to wear masks. Hence, the Japanese people found it difficult to speak through masks; thus, a better understanding of the hearing-impaired was gained. Moreover, in Tokyo's Minato Ward, the "Minato Ward Model" was launched, which is a program that was started on April 1, 2022, to subsidize the purchase of hearing aids for the elderly to support the early detection of hearing loss among this population, because this condition is one of the risk factors for dementia, and to enable the elderly to remain active and lively in the community [51]. Since then, this movement has been spreading gradually. Therefore, the current situation presents a slight improvement from the time of the survey.

However, people's understanding of hearing loss has not much improved yet. The result indicated that a lack of knowledge about hearing loss can lead to an inability to understand aged people with impaired hearing. To ensure the continued use of hearing aids by the individuals with hearing loss, specialists such as doctors, speech therapists, gerontologists, and other professionals should educate the people. Moreover, family members, local government officials, hospital staff, psychologists, and others with whom the elderly usually communicate must provide correct knowledge regarding this. According to a report by Morita et al., a volunteer organization in Denmark was introduced, showing the importance of such activities [47].

14.3 Suggestions for the Future

14.3.1 Educational and Collaborative Model

We propose a model to improve this situation and to enable early detection and intervention of age-related hearing loss. Based on the results of this study, the first step is to raise people's knowledge about hearing loss. The public has little knowledge about the characteristics of hearing loss, the relationship between hearing loss and other diseases, the importance of early detection and hearing replacement, the process of seeing a doctor and purchasing hearing aids after hearing loss occurs, and the adjustment of hearing aids. It is concerning that there is a tendency to use hearing aids without adjustment and to disseminate the perception of their uselessness in the community. Therefore, the awareness regarding hearing aids should be spread to the public. Senior citizens' clubs are organized throughout Japan. Consequently, it would be possible to educate and volunteer about hearing loss and hearing aids, as it is being done in Denmark [47]. A senior citizens' club would help prevent patients' isolation due to hearing loss. Additionally, rather than learning about hearing loss only at an older age, hearing health education should be included in the school curriculum. Learning about hearing health early in life will positively impact lifelong health.

Furthermore, not all medical professionals are necessarily hearing specialists, although otolaryngology, for example, covers the area from the skull to the neck. As previously mentioned, few otolaryngologists and speech-language pathologists specialize in hearing in Japan [49, 50]. A study conducted by Morita et al. revealed that hearing status was not assessed at admission, even in hospitals that admit many elderly patients [52]. They stated that to improve the quality of nursing care, it is essential for nurses to understand and evaluate patients' hearing accurately, including subjective and objective information, and to provide detailed support for patients who have hearing loss during their hospitalization [52]. In another study, Blazer stated that recognition and treatment of hearing loss should be the responsibility of all healthcare professionals working with older people, particularly psychiatrists and other mental healthcare workers, who are in a better position to determine the presence or absence of hearing loss [41]. Further, hearing loss can be detected when false answers to questions are obtained from an older adult undergoing a psychological evaluation.

Conversely, they often visit hospitals for physical ailments. Older people often report physical conditions, even when depressed. Hearing loss will likely be detected during a visit to another department since hearing loss has been associated with cardiovascular disease, diabetes, hypertension, falls, dementia, and depression [11, 14, 18, 19, 23–25, 28]. Therefore, healthcare professionals also need to have knowledge and skills regarding hearing loss, referrals, and hearing aids, as well as regarding communication with older adults with hearing loss.

Therefore, we propose an educational and collaborative model (Fig. 14.1). The squares in Fig. 14.1 represent facilities: municipalities/community support care centers, hospitals/clinics with audiology or speech and hearing specialists, other specialty hospitals (outpatient), other specialty hospitals (inpatient), hearing aid companies/providers, volunteer organizations/senior citizen clubs/peer support groups. Each square box, except for volunteers, is marked professional to the facility. The rounded squares indicate educational content related to age-related hearing loss. The left-hand square represents education for the general public, and the right-hand square represents education for professionals. Basic information on hearing loss, screening, hearing aids, etc., is standard. In addition, the instruction for the general public includes information on the benefits of hearing aids, how to purchase hearing aids, and grant programs for early intervention and continued hearing care. On the other hand, education for health professionals includes physical assessment, handling of hearing aids, battery replacement, and the relationship to other diseases. The black arrows indicate the implementation of the education. The central players here

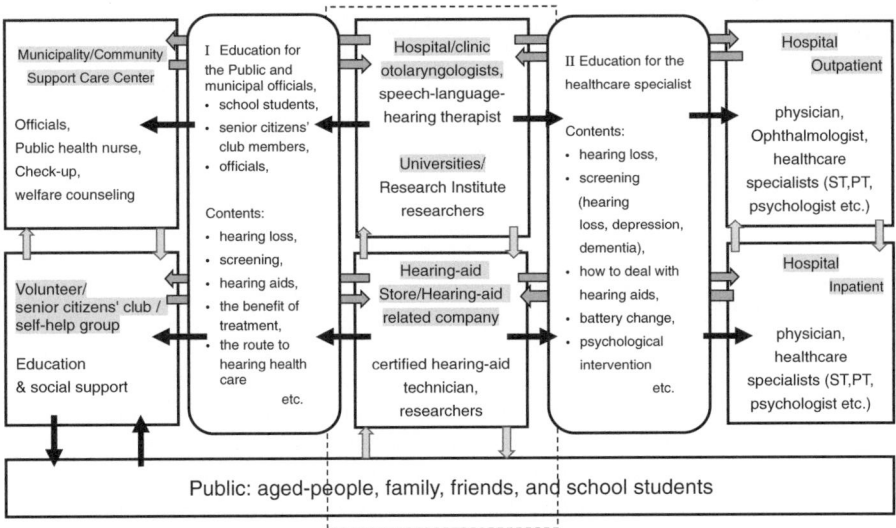

Fig. 14.1 Educational and collaborative model

are the audiology specialists and hearing aid providers in the center of the figure. These professionals educate the public, municipalities, volunteer organizations, and hospitals of other departments. Sano et al. (2021) introduce an Australian initiative, Hearing Australia, the most prominent national provider in Australia offering hearing and hearing aid education upon school and company request on-site [54]. Educated volunteers also educated older adults with hearing loss, as well as their families, friends, and school students. They also provide social support and prevent isolation. Hearing aid volunteers, such as those in Denmark [47], would be a good idea here.

The gray arrows in Fig. 14.1 indicate collaboration. When all medical professionals are well-informed about hearing loss and the facilities collaborate, they can detect it early, diagnose it appropriately, and provide the required treatment and care. This timely intervention can significantly improve the quality of life for the elderly, making it a matter of utmost importance. Hospitals could collaborate and refer those with hearing loss for early intervention. Audiologists are learning about psychological screening and contributing to mental health care in Australia, where audiologists are well established, because of the link between hearing loss and psychological challenges [42, 53]. If there were more speech-language pathologists or psychotherapists specializing in auditory care and clinical psychology in Japan, early intervention would be possible. They could refer the patient from otorhinolaryngology to psychiatry and vice versa, but currently, this is difficult. On the contrary, it would be more realistic in Japan for a physician in another department to notice a patient's hearing loss and refer them to an otolaryngologist or speech-language pathologist. Additionally, psychological support is necessary for successful hearing aid use. The cooperation of psychiatrists and clinical psychologists is also essential. The above education of health professionals and collaboration with various organizations will promote early detection and intervention of age-related hearing loss.

The question is how to establish such an educational system. The Australian approach may be helpful. For example, Macquarie University has a hearing hub where universities, clinics, providers, National Point of Services, and hearing aid equipment manufacturers collaborate on research, development, and treatment [54]. The first step is to create a virtual platform at a core university or other regional institution, collaborating with industry, government, and academia. Researchers share information there while developing and disseminating educational programs.

14.3.2 Screening Test

Some may think the medical field is too busy to screen for hearing loss. What if it were a simple, one- or two-second screening? What if it could be done in a typical living environment, with no need for a soundproof room or expensive audiometer, and with a certain degree of sensitivity and specificity? In this section, we introduce a simple screening test that allows anyone to screen for hearing loss in a simplified manner. Sano et al. (2018) studied the Finger rub/Finger tap screening test [55], referring to the method of Torres-Russotto et al. [56]. A "finger tap" was added to the conventional finger rub test, and a significant negative correlation was found between the hearing results and mean hearing acuity when the distance of the sound source varied from 5 cm, 30 cm, and 60 cm from the ear, confirming the concomitant validity of the test. The test was able to detect mild hearing loss greater than 26 dB with a sensitivity of 95.2% and specificity of 71.4% for the 5 cm and 30 cm conditions alone. Thus, age-related hearing loss makes it difficult to hear finger-rubbing sounds. Although there are individual differences in the sounds, finger rub sounds are thought to correspond to high frequencies and tap sounds to lower frequencies.

This test can detect more than mild hearing loss without shielding the non-tested ear. The test can be performed in normal daily sounds, in the absence of loud background noises like air conditioners and televisions. This test is very simple, noninvasive, excellent, and economical. These economical screening tests can be performed by anyone in any setting, including homes, hospitals, and institutions such as nursing homes, and require no special equipment. This test has the advantage of detecting a certain number of everyday sounds without requiring entry into a soundproof room. Thus, hearing loss, when suspected, is easily screened in a patient's room or examination room in another department.

Finger Rub/Finger Tap Test (Self-Check Version, Simplified Version Based on Sano et al. (2018)).

Preparation

1. Turn off the TV, radio, air conditioner, ventilation fan, and other sound-producing devices to ensure a quiet environment.
2. Wash your hands thoroughly and wipe them dry with a dry towel to keep your fingers dry.

1. Finger rub test (Fig. 14.2).

   - Rub the side of the index finger lightly and quickly in an arc with the thumb.
   - Rub your fingers together about 4–5 times quickly, 5 cm from the ear.

Fig. 14.2 Finger rub/finger tap screening test (simplified version based on Sano et al. (2018))

2. Finger Tap Test (Fig. 14.2).

- The tap should be made by lightly aligning the thumb and index finger.
- Light tapping by the thumb and index finger together is done. At this time, there is no need to try to make a particular sound. Just match them up. Be careful not to make a snapping sound.
- Make finger tap sounds about 4–5 times, 5 cm from the ear.

If you cannot hear at least once, you may have a hearing loss. Hence, consult an otolaryngologist.

14.4 Conclusion

Age-related hearing loss is one of the most common chronic conditions; however, many people do not immediately see a doctor when they notice hearing loss. Untreated hearing loss can affect older adults' health and quality of life. In this chapter, we first described age-related hearing loss, the physical and psychosocial effects, and how to cope with hearing loss, including the use of hearing aids and communication strategies. Following this, we discussed the difficulties of early detection and intervention from the viewpoint of the Japanese social system and a standard of care that is not well known. Thirdly, we discussed the results of our survey conducted on local government officials throughout Japan, which revealed that few local governments in Japan conduct hearing tests for the elderly. At the time of the survey, only two percent of the municipalities that responded had implemented it. And that some local government officials did not know enough about hearing loss or how to deal with people with hearing difficulties. The staff members with more knowledge can more appropriately respond to elderly persons with hearing loss. Further, groups with more knowledge were more willing to learn more about hearing loss and wear hearing aids when they had hearing loss, suggesting that they had a low prejudice toward hearing aids. Based on the above, we proposed an educational and collaborative model that enables the early detection of hearing loss and intervention to help those with hearing difficulties. In this proposed model, municipalities, researchers, and physicians will work together to educate the general public, other municipalities, and schools to spread knowledge about hearing loss and hearing aids. In addition, local government officials and elderly members of local senior citizens' clubs who have received special education will take the role of spreading knowledge to the next step. Medical institutions are not to be divided vertically, but need to strengthen horizontal cooperation with local governments, hospitals and other relevant institutions to achieve early detection and continuous care. Finally, we introduce the Finger Rub/Finger Tap screening test for the early detection of hearing loss. This is a test that anyone can quickly and efficiently perform.

Acknowledgments This research was supported by Grant-in-Aid for Scientific Research(C), (19K02219, Tomoko Sano).

The authors would like to thank Enago for the English language review.

References

1. World Health Organization. Deafness and hearing loss. https://www.who.int/health-topics/hearing-loss#tab=tab_2. Accessed 31 Jan 2023.
2. Uchida Y, Sugiura S, Nakashima T, Ando F, Shimokata H. Estimates of the size of the hearing-impaired elderly population in Japan and 10-year incidence of hearing loss by age, based on data from the National Institute for longevity sciences-Longitudiual study of aging (NILS-LSA). Nihon Ronen Igakkai Zasshi. Jpn J Geriatr. 2012;49:222–7. https://doi.org/10.3143/geriatrics.49.222.
3. Statistics Bureau of Japan. https://www.stat.go.jp/data/topics/topi1321.html. 2022. Accessed 31 Jan 2023.
4. Cabinet Office. White paper on aging. 2020. https://www8.cao.go.jp/kourei/whitepaper/w-2020/zenbun/pdf/1s1s_02.pdf. Accessed 31 Jan 2023.
5. Bowl MR, Dawson SJ. Age-related hearing loss. Cold Spring Harb Perspect Med. 2019;9(8):a033217. Published 2019 Aug 1. https://doi.org/10.1101/cshperspect.a033217.
6. Gates GA, Mills JH. Presbycusis. Lancet. 2005;2005(366):1111–20.
7. Fransen E, Lemkens N, Van Laer L, Van Camp G. Age-related hearing impairment (ARHI): environmental risk factors and genetic prospects. Exp Gerontol. 2003;38(4):353–9. https://doi.org/10.1016/s0531-5565(03)00032-9.
8. Wong AC, Ryan AF. Mechanisms of sensorineural cell damage, death and survival in the cochlea. Front Aging Neurosci. 2015;7:58. https://doi.org/10.3389/fnagi.2015.00058.
9. Mader S. Hearing impairment in elderly persons. J Am Geriatr Soc. 1984;32(7):548–53. https://doi.org/10.1111/j.1532-5415.1984.tb02245.x.
10. Slawinski EB, Hartel DM, Kline DW. Self-reported hearing problems in daily life throughout adulthood. Psychol Aging. 1993;8(4):552–61. https://doi.org/10.1037/0882-7974.8.4.552.
11. McKee MM, Stransky ML, Reichard A. Hearing loss and associated medical conditions among individuals 65 years and older. Disabil Health J. 2018;11(1):122–5. https://doi.org/10.1016/j.dhjo.2017.05.007.
12. Chopra A, Jung M, Kaplan RC, Appel DW, Dinces EA, Dhar S, et al. Sleep apnea is associated with hearing impairment: the Hispanic community health study/study of Latinos. J Clin Sleep Med. 2016;12(5):719–26. https://doi.org/10.5664/jcsm.5804.
13. Nakayama M. How do sleep disturbance affect inner ear? Nagoya Med J. 2013;53:115–21.
14. Dalton DS, Cruickshanks KJ, Klein BE, Klein R, Wiley TL, Nondahl DM. The impact of hearing loss on quality of life in older adults. Gerontologist. 2003;43(5):661–8. https://doi.org/10.1093/geront/43.5.661.
15. Ventry IM, Weinstein BE. The hearing handicap inventory for the elderly: a new tool. Ear Hear. 1982;3(3):128–34. https://doi.org/10.1097/00003446-198205000-00006.
16. Gopinath B, Schneider J, McMahon CM, Teber E, Leeder SR, Mitchell P. Severity of age-related hearing loss is associated with impaired activities of daily living. Age Ageing. 2012;41(2):195–200. https://doi.org/10.1093/ageing/afr155.
17. Hyams AV, Hay-McCutcheon M, Scogin F. Hearing and quality of life in older adults. J Clin Psychol. 2018;74(10):1874–83. https://doi.org/10.1002/jclp.22648.
18. Viljanen A, Kaprio J, Pyykkö I, Sorri M, Pajala S, Kauppinen M, et al. Hearing as a predictor of falls and postural balance in older female twins. J Gerontol A Biol Sci Med Sci. 2009;64(2):312–7. https://doi.org/10.1093/gerona/gln015.

19. Lin FR, Ferrucci L. Hearing loss and falls among older adults in the United States. Arch Intern Med. 2012;172(4):369–71. https://doi.org/10.1001/archinternmed.2011.728.
20. Campbell J, Sharma A. Cross-modal re-organization in adults with early stage hearing loss. PLoS One. 2014;9(2):e90594. https://doi.org/10.1371/journal.pone.0090594.
21. Glick H, Sharma A. Cross-modal plasticity in developmental and age-related hearing loss: clinical implications. Hear Res. 2017;343:191–201. https://doi.org/10.1016/j.heares.2016.08.012.
22. Cardon G, Sharma A. Somatosensory cross-modal reorganization in adults with age-related, early-stage hearing loss. Front Hum Neurosci. 2018;12:172. https://doi.org/10.3389/fnhum.2018.00172.
23. Herbst KG, Humphrey C. Hearing impairment and mental state in the elderly living at home. Br Med J. 1980;281(6245):903–5. https://doi.org/10.1136/bmj.281.6245.903.
24. Sumi E, Takechi H, Wada T, Ishine M, Wakatsuki Y, Murayama T, et al. Comprehensive geriatric assessment for outpatients is important for the detection of functional disabilities and depressive symptoms associated with sensory impairment as well as for the screening of cognitive impairment. Geriat Gerontol Int. 2006;6:94–100.
25. Cosh S, Helmer C, Delcourt C, Robins TG, Tully PJ. Depression in elderly patients with hearing loss: current perspectives. Clin Interv Aging. 2019;14:1471–80. https://doi.org/10.2147/CIA.S195824.
26. Skelton D. Hearing impairment in the elderly. Can Fam Physician. 1984;30:611–5.
27. Jayakody DMP, Wishart J, Stegeman I, Eikelboom R, Moyle TC, Yiannos JM, et al. Is there an association between untreated hearing loss and psychosocial outcomes? Front Aging Neurosci. 2022;14:868673. https://doi.org/10.3389/fnagi.2022.868673.
28. Saito H, Nishiwaki Y, Michikawa T, Kikuchi Y, Mizutari K, Takebayashi T, et al. Hearing handicap predicts the development of depressive symptoms after 3 years in older community-dwelling Japanese. J Am Geriatr Soc. 2010;58(1):93–7. https://doi.org/10.1111/j.1532-5415.2009.02615.x.
29. Uhlman FR, Larson EB, Koepsell TD. Hearing impairment and cognitive decline in senile dementia of the Alzheimer's type. J Am Geriatr Soc. 1986;34(3):207–10. https://doi.org/10.1111/j.1532-5415.1986.tb04204.x.
30. Lin FR, Metter EJ, O'Brien RJ, Resnick SM, Zonderman AB, Ferrucci L. Hearing loss and incident dementia. Arch Neurol. 2011;68(2):214–20. https://doi.org/10.1001/archneurol.2010.362.
31. Livingston G, Huntley J, Sommerlad A, Ames D, Ballard C, Banerjee S, et al. Dementia prevention, intervention, and care: 2020 report of the lancet commission. Lancet. 2020;396(10248):413–46. https://doi.org/10.1016/S0140-6736(20)30367-6.
32. Stahl SM. Does treating hearing loss prevent or slow the progress of dementia? Hearing is not all in the ears, but who's listening? CNS Spectr. 2017;22(3):247–50. https://doi.org/10.1017/S1092852917000268.
33. Uchida Y, Sugiura S, Nishita Y, Saji N, Sone M, Ueda H. Age-related hearing loss and cognitive decline—the potential mechanisms linking the two. Auris Nasus Larynx. 2019;46(1):1–9. https://doi.org/10.1016/j.anl.2018.08.010.
34. Wayne RV, Johnsrude IS. A review of causal mechanisms underlying the link between age-related hearing loss and cognitive decline. Ageing Res Rev. 2015;23(Pt B):154–66. https://doi.org/10.1016/j.arr.2015.06.002.
35. Amieva H, Ouvrard C, Meillon C, Rullier L, Dartigues JF. Death, depression, disability, and dementia associated with self-reported hearing problems: a 25-year study. J Gerontol A Biol Sci Med Sci. 2018;73(10):1383–9. https://doi.org/10.1093/gerona/glx250.
36. Boi R, Racca L, Cavallero A, Carpaneto V, Racca M, Dall' Acqua F, et al. Hearing loss and depressive symptoms in elderly patients. Geriatr Gerontol Int. 2012;12(3):440–5. https://doi.org/10.1111/j.1447-0594.2011.00789.x.
37. McConnell EA. Communicating with a hearing-impaired patient. Nursing. 1998;28(1):32. https://doi.org/10.1097/00152193-199801000-00019.
38. McConnell EA. How to converse with a hearing-impaired patient. Nursing. 2002;32(8):20. https://doi.org/10.1097/00152193-200208000-00017.

39. Dugan MB. Living with hearing loss. Washington: Gallaudet University Press; 2003.
40. Saito Y, Yajima Y. The effects on mental health of coping strategies for hearing loss in elderly. J Jap Aca Health Sci. 2005;8(2):89–97.
41. Blazer DG. Hearing loss: the silent risk for psychiatric disorders in late life. Psychiatr Clin North Am. 2018;41(1):19–27. https://doi.org/10.1016/j.psc.2017.10.002.
42. Bennett RJ, Meyer CJ, Ryan B, Barr C, Laird E, Eikelboom RH. Knowledge, beliefs, and practices of Australian audiologists in addressing the mental health needs of adults with hearing loss. Am J Audiol. 2020;29(2):129–42. https://doi.org/10.1044/2019_AJA-19-00087.
43. Brooks DN. Hearing aid candidates—some relevant features. Br J Audiol. 1979;13(3):81–4. https://doi.org/10.3109/03005367909078882.
44. Davis A, Smith P. Adult hearing screening: health policy issues—what happens next? Am J Audiol. 2013;22(1):167–70. https://doi.org/10.1044/1059-0889(2013/12-0062).
45. Uchida Y, Nakashima T, Ando F, Niino N, Shimokata H. Prevalence of self-perceived auditory problems and their relation to audiometric thresholds in a middle-aged to elderly population. Acta Otolaryngol. 2003;123(5):618–26. https://doi.org/10.1080/00016480310001448.
46. David D, Zoizner G, Werner P. Self-stigma and age-related hearing loss: a qualitative study of stigma formation and dimensions. Am J Audiol. 2018;27(1):126–36. https://doi.org/10.1044/2017_AJA-17-0050.
47. Morita K, Sano T, Sugizaki K, Ito N, Okuyama Y. Danish approach to hearing impairment of the elderly and challenges facing Japan: significance of the peer support by senior hearing volunteers. Appl Gerontol. 2022;16(1):119–127.45.
48. Japan Trak. Report. 2022. http://www.hochouki.com/files/JAPAN_Trak_2022_report.pdf. Accessed 8 Feb 2023.
49. Speech-Language-Hearing Therapists: Japanese Association of Speech-Language-Hearing Therapists. https://www.japanslht.or.jp/what/. 2022. Accessed 31 Jan 2023.
50. Hearing aids consultation Doctor: Japanese Society of Otorhinolaryngology Head and Neck Surgery. https://www.jibika.or.jp/modules/certification/index.php?content_id=39. 2022. Accessed 31 Jan 2023.
51. Subsidy program to purchase hearing aids for the elderly: Minato Ward. https://www.city.minato.tokyo.jp/zaitakushien/koureisya/hoctyouki/hotyoukijyosei.html. 2022. Accessed 31 Jan 2023.
52. Morita K, Ito N, Sano T, Furuta A, Fujisaki S, Osada H. Evaluation of hearing function in the elderly: analyses of the current status by nurses of the general hospitals. J Gerontol Res. 2014;4:67–80.
53. Bennett RJ, Kelsall-Foreman I, Donaldson S, Olaithe M, Saulsman L, Badcock JC. Exploring current practice, knowledge, and training needs for managing psychosocial concerns in the audiology setting: perspectives of audiologists, audiology reception staff, and managers. Am J Audiol. 2021;30(3):557–89. https://doi.org/10.1044/2021_AJA-20-00189.
54. Sano T, Morita K, Sugizaki K. Challenges for supporting those with hearing loss in Japan: an introduction of the Australian support system for hearing loss and the importance of industry-government-academia-private sector collaboration. Appl Gerontol. 2021;15(1):86–96.
55. Sano T, Morita K, Okuyama Y, Ito N, Osada H. Examining the validity of the finger rub/finger tap screening test for the early detection of age-related hearing loss. Jpn J Public Health. 2018;65(6):288–99.
56. Torres-Russotto D, Landau WM, Harding GW, Bohne BA, Sun K, Sinatra PM. Calibrated finger rub auditory screening test (CALFRAST). Neurology. 2009;72(18):1595–600. https://doi.org/10.1212/WNL.0b013e3181a41280.

Chapter 15
The Role of Aesthetics in Elderly Care

Noriko Onishi and Toshiya Nagamatsu

Abstract In Japan's super-aged society, there is an ongoing debate as to whether aesthetics can be used, along with medical and nursing care, as a means to support the health of elderly people. Some people understand aesthetics as the act of beautifying one's appearance (e.g., using make-up) rather than maintaining one's personal hygiene.

It is widely known that external beauty has often been influenced by the collective values and the cultural and social trends of the time, and, in some cases, make-up and clothing have been regarded as detrimental to health.

However, in a society where the ageing rate has been increasing so rapidly, it is no longer possible to face future challenges, such as the extension of healthy life expectancy, as advocated by the WHO, through medical and nursing care alone.

Moreover, the meaning of health in societies enjoying longevity is shifting from the physical to the psychological and social aspects that may allow older people to enjoy a lively and fulfilling existence, and aesthetics, in this sense, is gaining recognition as a means to support health.

In order to extend healthy life expectancy, it is important for the elderly to actively engage with society and prevent a decline in their level of independence, and, in this respect, aesthetics can play a key role in maintaining their involvement with local communities.

Therefore, in this paper, we introduce a study on cosmetic interventions aimed at preventing dementia and frailty, two health issues often encountered in super-aged societies, and assess the role of aesthetics as a means to support health in elderly people. In addition, the authors introduce a specific area of study called "aesthetics and welfare," which involves training beauty practitioners to extend

---

N. Onishi (✉) · T. Nagamatsu
Yamano College of Aesthetics, Tokyo, Japan
e-mail: nonishi@skycampus.jp; toshiya.nagamatsu@skycampus.jp

healthy life expectancy. The authors propose a multifactorial intervention programme, utilising aesthetics as a means to promote social activities and interaction for the elderly.

Keywords Medical and nursing care of the elderly · Healthy life expectancy · Dementia · Frailty aesthetics · Aesthetics and welfare

15.1 Aesthetics and Public Health in Japan

C.E.A. Winslow defined public health as the "science and art of preventing disease, prolonging life, and promoting physical and mental health and efficiency through organised community efforts" [1].

On the other hand, hairstylists and beauticians in Japan provide services "contributing to the improvement of public health," as specified in Article 1 of the Barbers Act and Cosmetologists Act [2]. However, the relationship between aesthetic and physical/mental health is generally poorly understood, and many professionals have been traditionally quite dismissive of it, especially in the field of medical and nursing care for the elderly. One reason for this is that the term "aesthetic"is perceived as the act of beautifying one's appearance (e.g., using make-up) rather than the act of preserving one's personal hygiene (e.g., washing one's face or hair). Sometimes beautifying one's appearance can be harmful both physically and mentally, because of the influence exerted by the values of the community to which individuals belong (e.g., the time, cultural and social trends). In medical and nursing care settings, the notion of beautifying has long been neglected because of the difficulties in obtaining bodily information such as facial complexion and because it is thought to increase the burden on caregivers.

However, in Japan's super-aged society, it has become important to promote "healthy ageing," as advocated by the WHO. For example, maintaining as much independence as possible, even in old age, and prolonging the time for one to enjoy a vibrant life experience may help reduce social burdens in the future.

It is worth noting that problems related to the extension of healthy life expectancy cannot be solved by a healthcare system that promotes medical and nursing care only when needed. In this respect, the creation of a system enabling local residents to actively promote the extension of healthy life expectancy is increasingly needed. The authors wish to incorporate an aesthetic element into a system that promotes collaboration with other providers involved in the prevention and improvement of dementia and frailty.

Intervention studies focusing on dementia and frailty, which are causes of reduced independence, are presented later. In some of these studies, expressions such as "cosmetics"or "cosmetic therapy"are used. This is because the interventions involve the use of make-up, although in many cases, health outcomes in older people result from multiple factors. In other words, in addition to make-up, haircut and clothing, the words and behaviour of beauticians and the people around them, all play a significant part in achieving the final result. Therefore, the authors decided to include all these factors under the term "aesthetic." Moreover, the health outcomes of these interventions are greatly influenced by social interaction and participation rather than only by cosmetic behaviour. While cosmetic behaviour may act as a catalyst in these multifactorial interventions, the health outcomes obtained are largely due to a multiplicity of factors.

15.2 Elderly Care and Aesthetics

In Japan, the nursing care insurance system was launched in 2000 to solve the social issues of a super-aged society. The system is founded on the central tenet that the whole community is responsible for the welfare of the elderly.

The notion of "aesthetics and welfare" was proposed by Masayoshi Yamano of the Yamano Academy in 1999. The expression was coined with the hope that combining two such areas, aesthetics and welfare, through interdisciplinary and practical education would contribute to the future well-being of society.

In addition to promoting education on aesthetics and welfare, the Japanese Society of Welfare and Aesthetics was established to promote research focusing on this discipline and train learners as "aesthetics and welfare professionals." In order to improve the quality of these professionals, the NPO National Association for Care, Welfare and Aesthetics was established in 2002 to provide practical training.

In the early 2000s, there was no formal recognition of the role of aesthetics in increasing the well-being of older people in the field of medical and nursing care for the elderly in Japan. Although the cooperative mechanisms have yet to be refined, the importance of aesthetics, at least in the aforementioned fields, has been recognised. This is partly due to the long-term contributions of aesthetics to society through the activities of beauty companies as well as our Academy and partly due to the increasing number of studies in psychology, medicine, and nursing [3, 4]. A representative nursing study conducted in 1993 focusing on a hospital-business partnership showed that cosmetic interventions may improve cognitive function and independence in older people affected by dementia. This and other studies are discussed below.

Aesthetics and welfare recommended the provision of cosmetic services to older people in need of nursing care, who account for 28.7% of the general population (statistics from the Ministry of Internal Affairs and Communications, 2020), which is approximately 20% of our country's elderly population. However, there is a concern that the remaining 80% would also lose their independence if they were to lead an inactive life. On top of that, containment measures against the COVID-19 pandemic, such as activity restrictions, further reduced older people's independence. One of the future academic challenges in the field of aesthetics and welfare is to verify whether aesthetics can effectively contribute to preventing the decline of independence in the remaining 80% of the elderly population.

15.3 Dementia and Aesthetics

The acknowledgement of aesthetics as an effective measure in medical and nursing care is largely due to a cosmetic intervention study [5] conducted in 1993 at Naruto Yamakami Hospital on 40 elderly patients (aged 66–93) affected by dementia. By the end of the study, 11 out of 40 participants no longer needed to wear a diaper.

In Japan, at that time, older people did not pay particular attention to personal grooming or make-up, and it was common for all women staying in hospitals and nursing homes to have the same short hairstyle and no make-up. The cosmetic intervention was carried out in this historical context. The intervention setting must have provided quite an unusual scene in which a smiling cosmetic consultant came along to apply make-up on the patients while they cheerfully smiled and chatted with the hospital staff members. As a result of the intervention, the facial expressions of a large number of patients ($n = 36$) became livelier. The intervention also improved not only the communication between the staff members and some patients but also the effectiveness of rehabilitation ($n = 11$). It can be expected that as older people adopt a more active lifestyle, and the number of times they can reach the toilet by walking increases, some may no longer need to wear a diaper (Fig. 15.1).

In 2013, an ongoing cosmetic intervention study for the prevention of dementia was launched by Ikeyama et al [6] Daily skincare was continuously performed for a 3-month period on 16 women living in an elderly facility, resulting in the improvement of their motor-FIM (Functional Independence Measure) items. Furthermore, the EEG pattern (Dα) of six patients, who participated in make-up classes twice a month for 6 months, was stabilised. The assessment of independence in ADL (Activity of Daily Living), using FIM, showed improvement, especially related to cognitive items (e.g., social interaction, problem-solving, and memory). It is

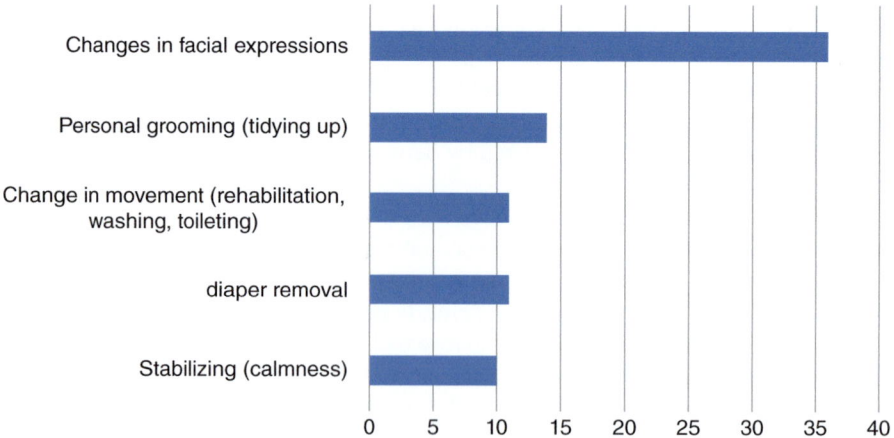

Fig. 15.1 Naruto Yamakami Hospital ($n = 40$) Elderly people with dementia aged 66–93. Multiple responses (1993)

expected that positive social interaction between the patients and the people involved in this intervention study might have continued, along with the practice of skincare and make-up routines. In other words, as mentioned above, the consistent practice of skincare and make-up routines sparked an interest in hairstyle and clothing and increased personal grooming activities. The positive communication opportunities with the people around them are also considered to be an outcome of the intervention.

A study conducted by Ohsugi et al. in 2015 [7] compared changes in 20 elderly women requiring nursing care who received weekly cosmetic interventions for 3 months with 13 women who did not receive any intervention. As a result, the intervention group was shown to have improved cognitive and psychic functions and reduced decline in bodily function. Ohsugi et al. believed that multiple factors at play resulted in these outcomes, including the ability to put on make-up, which was previously not possible, the therapist's deeper involvement, and more frequent face washing and personal grooming activities.

In other words, multifactorial interventions that include such environmental factors, along with the use of make-up, are important to prevent the decline of cognitive function. Therefore, it is important for professionals working in the medical and nursing care sector and other related fields to collaborate on a multi-disciplinary level in order to consciously create an environment wherein this plurality of factors may intervene.

15.4 Frailty and Aesthetics

Healthy life expectancy, a new indicator proposed by the WHO, refers to the average number of years that a person can expect to live in full health without requiring long-term care for such disabling conditions as dementia or being bedridden (in 2018 the average life expectancy of Japanese people was 87.32 years for women and 81.25 years for men, while healthy life expectancy was 74.79 years for women and 72.14 for men).

Frailty, which can be defined as an intermediate condition between health and the need for care (such as in the case of bedridden patients), refers to the deterioration of physical and mental vitality (e.g., muscle strength, cognitive functioning, social interaction, etc.) associated with old age. However, if signs of this condition are detected early and appropriate corrective measures are taken (e.g., reconsidering one's own daily lifestyle), it is possible to control its progression and return to a healthy state.

The vicious circle generated by this condition is, for example, that as people grow older and have fewer opportunities to go out and meet other people, their physical fitness and cognitive functions tend to deteriorate, and they become even more reluctant to go out and engage in social interactions. Moreover, if older people do not go out, their appetite may reduce, and their diet may become unbalanced if they do not go shopping. As the performance of activities of daily living declines, under-utilised physical and mental functions deteriorate, leading to an increasingly low level of independence and the need for care.

To break this vicious circle, the Ministry of Health, Labour and Welfare recommended "food," "physical activity," and "social interaction." In order to build consistent opportunities for "social interaction," it has been recommended the creation of "kayoi-no-ba" (literally meaning a "place for regular visits"), as support projects for community-based care prevention activities.

According to Document 2-1 of the third Meeting on Promotion Policies for General Care Prevention Services held in 2019, the term "kayoi-no-ba" refers to the "regional development of activities conducive to care prevention in which any older person can participate, irrespective of age, physical or mental conditions, etc.". Such activities, which take place in resident-led community salons (i.e., "kayoi-no-ba"), are actively sponsored by local governments in accordance with local conditions."

In this respect, local beauty salons, which are easily accessible, and provide a place for conversing and socialising, may fulfill this role. As such, entire communities can be expected to form around beauty salons.

However, research findings show that even in 2017, before the outbreak of the COVID-19 pandemic, only 4.9% of the elderly population had access to "kayoi-no-ba."

The main goal of the Ministry of Health, Labour and Welfare by 2025, when the baby-boom generation will be 75 years of age or older, is to develop a community-based integrated care system that can provide integrated housing, medical/nursing care, and prevention/livelihood support so that people can go on living where they feel most comfortable, until the end of their lives, even if they are in need of serious long-term care. Therefore, it is thought that a more efficient community-based integrated care system can be created if hairdressing/beauty salons, which are easily accessible, regularly attended by older people, and where friendly relationships can be established, could be used as "kayoi-no-ba" to develop a partnership system with experts in the fields of medical and nursing care (Fig. 15.2).

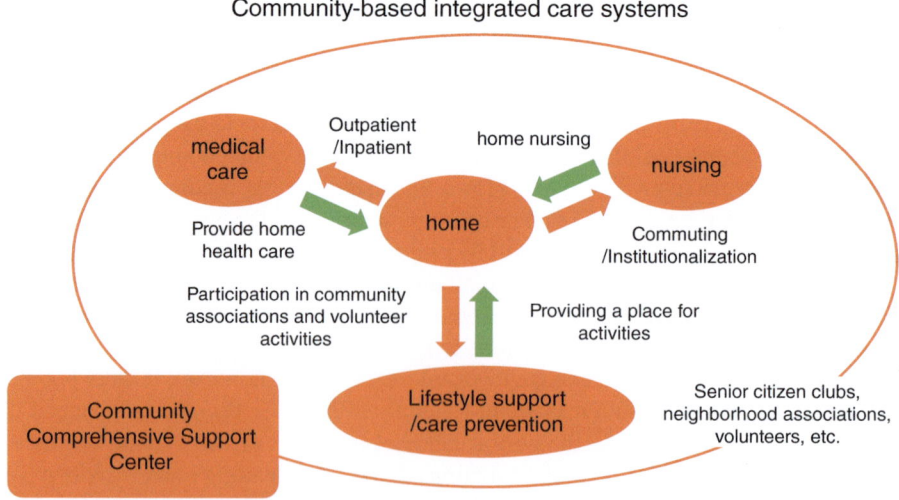

Fig. 15.2 Modified figure from the community-based integrated care workshop report March 2016

15.5 Aesthetics for Extending Healthy Life Expectancy

Another cosmetic intervention study (Kawai et al. 2016) [8], conducted in 2014 by the Tokyo Metropolitan Geriatric Hospital and Institute of Gerontology and others, measured the effects of the intervention on the subjective health views of community-dwelling older people, as well as their frequency of outing, and depressive tendencies. The intervention group that underwent the bi-monthly make-up programme for 3 months was compared to the control group that did not receive any intervention. The results showed that people who participated in the programme maintained or improved their subjective health views and depressive tendencies, while the frequency of outings for the control group decreased significantly.

Beauty salons have started to provide home beauty services to elderly people who cannot go out. Obuchi, who was involved in the study, later stated that, as a result of the make-up programme, all the participants in the intervention group were unanimously praised by people around them, with remarks such as, "You look prettier!" or "You look a bit different!" In particular, Obuchi pointed out that the improvement in subjective health views and depressive tendencies, as well as a renewed interest in regular outings observed in the study's participants, were the result not only of simply "applying make-up" but also of satisfying the need for social interaction and approval through the use of make-up, as exemplified by the unanimous praise received from the people around them. These are precisely the effects of multifactorial interventions that are mentioned above.

Abe describes the act of applying make-up as "an ambivalent act that can be both a nuisance and a diversion, depending on individual characteristics, the social context in which it takes place, and the circumstances that precede and follow it" [9]. This means that make-up may prove burdensome for some people, depending on their relationship with society and others. Therefore, it can be said that multifactorial interventions may take on diverse forms.

This is also something to keep in mind when conducting various cosmetic intervention studies. This is because, for example, the effect of wearing make-up is influenced by factors such as individual characteristics and relationships with other people. In fact, while some people may react positively when they are told that they look prettier after wearing make-up, others who are less confident in their looks may only focus on hiding what they perceive as a facial flaw and might not feel inclined to enjoy social interactions. This may lead some people to become obsessed with wearing make-up for the sole purpose of socialising. In such cases, the use of make-up may prove detrimental to one's health. In this respect, a person who suffered from anthropophobia stated the following: "The day I got to meet people without wearing make-up, I started to accept myself the way I was, enjoyed interacting with others, and felt that I had overcome my illness." In other words, beauty providers should be aware that the meaning of aesthetics may vary according to individual characteristics and the social context that precedes and follows it.

15.6 Conclusion

In Finland, a successful multifactorial intervention study, the Finnish Intervention Study for the Prevention of Cognitive Impairment by Lifestyle Interventions in the Elderly (FINGER) (FINnish GERiatrics) trial, was conducted from 2009 to 2011. The same methodology as the FINGER trial and will focus mainly on diet, physical activity and various activities that may help stimulate cognitive function.

The authors aim to create a multifactorial intervention programme that may help maintain cognitive function and prevent frailty, utilising aesthetics to promote social activities and interaction in daily life.

We believe healthy aging can only be achieved in a thriving and joyful community.

Until now, medical care and long-term care in Japan have focused on treating and preventing physical and mental illness. However, in a super-aged society, the value should be placed not on treatment and prevention but rather on a full and joyful life, even if one has a chronic disease or a disability from which recovery is not possible. This is why cosmetic intervention is so important. However, what can be demonstrated as an effect of a cosmetic intervention alone has been a challenge. As described in this paper, beauty does not lead to health effects by itself but is always accompanied by some multifactorial intervention. Beauty is also a continuous intervention in people's lives for at least 3 months, leading to the results. We always questioned the validity of the results when we said that beauty was effective. What we have learned is that beauty interventions rather facilitate multifactorial interventions. In other words, beauty has a role in facilitating multifactorial interventions for individuals and populations.

We intend to continue our research on beauty intervention as an important element in realizing a society aimed at by gerontology.

References

1. Winslow CEA. The untilled fields of public health. Science. 1920;51(1306):23.
2. Barbers Act (Act no. 234 of 1947), Cosmetologists Act (Act no. 163 of 1957).
3. Nozawa K, Sawazaki T. Review of the study on clinical psychological effects of cosmetic techniques. Mejiro J Psychol. 2006;2:49–63.
4. Cardenas X, Nishio Y, Fukui N, Tanaka K, Morikawa S, Suehara K. Article review about the application of make-up therapy in Japanese clinical site. Osaka Med Coll J Nurs Res. 2013;3:69–77.
5. Hayami M. Make-up as a means to regain one's own independence in daily life: an experimental trial conducted at Naruto Yamakami hospital. Month Gen Care. 1994;4(11):27–34.
6. Ikeyama K, Machida A, Shirato M, Takada S. Cosmetic behaviour as a tool for dementia prevention. In: Proceedings of the 3rd annual meeting of the Japan Society for Dementia Prevention, vol. 134; 2013.

7. Ohsugi H, Murata S, Murata J, Haruki K, Sokabe M, Oyama M. Tomiko Tani effects of intervention of make-up for frail elderly women from the viewpoints of physical, cognitive, and mental function. J Jpn Soc Early Stage Dement. 2015;8(2):71–7.
8. Kawai H, Inomata T, Otsuka R, Sugiyama Y, Hirano H. Shuichi Obuchi the effectiveness of beauty care on self-rated health among community-dwelling older people. Jpn J Geriat. 2016;53(2):123–32.
9. Abe T. The social psychophysiology of stress and cosmetic behaviour. Fragr J. 2006.

Chapter 16
Maintenance and Improvement of Brain Health by the Brainwave-Based Brain-Training Competition, "bSports"

Ryohei P. Hasegawa

Abstract This research and development aims to maintain and improve the brain health of older people and people with disabilities who have impaired motor function by repeatedly participating in cognitive training competitions known as "bSports" with real-time decoding of brainwave data and their feedback.

Keywords Brain-machine/computer Interface · Event-related potential · Communication aid · Cognitive assessment · Cognitive training

16.1 Introduction

In recent years, in preventive medicine, the importance of early detection and treatment of "frailty," the mental and physical state immediately before the need for nursing care, has been highlighted, and its recognition is spreading [1]. Dementia, which is on the rise mainly in developed countries, is receiving particular attention in this field. According to the World Alzheimer Report 2021, there are already more than 55 million patients with Alzheimer's disease worldwide, with enormous

---

R. P. Hasegawa (✉)
Human Augmentation Research Center (HARC), National Institute of Advanced Industrial Science and Technology (AIST), Tsukuba, Ibaraki, Japan

Faculty of Engineering, University of Fukui, Fukui, Fukui, Japan

Innovative Research Center for Preventive Medical Engineering, Nagoya University, Nagoya, Aichi, Japan

Parallel Brain Interaction Sensing Division, Tokyo University of Science, Chiba, Noda, Japan
e-mail: r-hasegawa@aist.go.jp

economic losses of approximately $1 trillion [2]. The cognitive function of dementia patients gradually declines due to various causes, interfering with daily life as symptoms progress. Another major issue is that older individuals with latent dementia are prone to serious accidents while driving. Although no cure for dementia has been established yet, mild cognitive impairment (MCI) occurs as a possible pre-stage of dementia [3]. Approximately 40% of patients recover from MCI without cognitive decline [4]. Therefore, it is important to detect MCI as early as possible and perform appropriate early interventions.

"Cognitive training" is attracting attention as a candidate for preventive interventions in healthy older individuals and patients with MCI [5]. Cognitive training is a program based on certain procedures intended to train and improve cognitive abilities, such as attention and memory, necessary for daily life. A large-scale study of healthy participants and patients with MCI also showed the contribution of cognitive training to the improvement of executive function [6]. In Japan, computer games (so-called "brain training") that can be experienced on home-use game consoles are widely popular among the general public. The WHO guidelines for dementia prevention highlight the importance of social activities and cognitive training [7] (Fig. 16.1).

On the other hand, older individuals with a natural decline in motor function and people with motor disabilities cannot play brain-training games as effective as healthy people. The "brain-machine/computer interface (BMI/BCI)," which directly connects the brain and machine [8, 9], could potentially solve this problem. In recent years, it has become possible for BMI users to operate external devices, such as personal computers, directly with brain activity, without words or gestures.

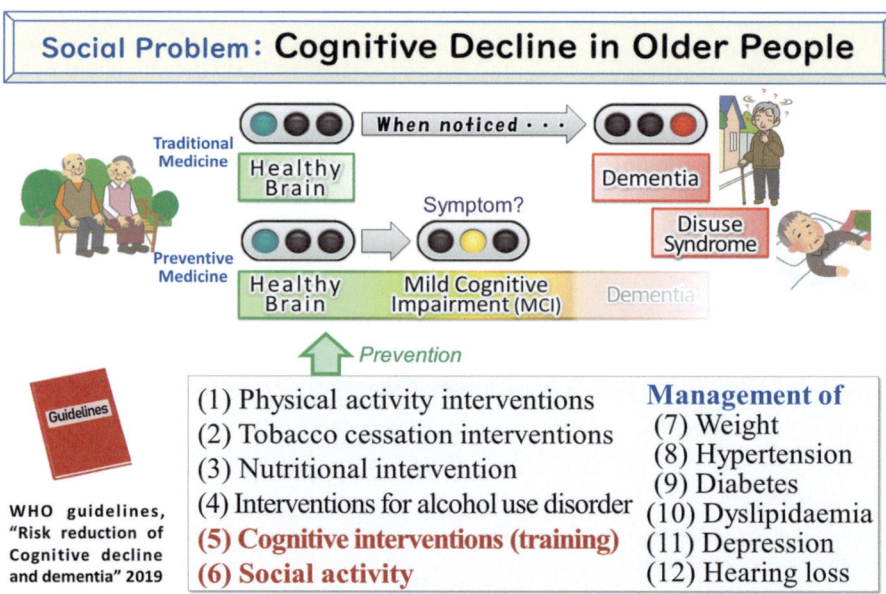

Fig. 16.1 Cognitive training and social activity may reduce the risk of cognitive decline

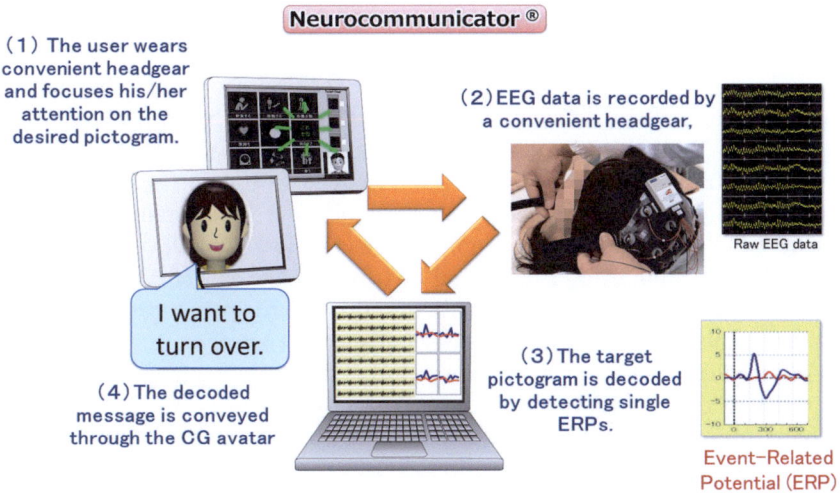

Fig. 16.2 ERP-based BMI, "Neurocommunicator" that uses the EEG Switch

Electroencephalogram (EEG)-based BMI, which measures the brain-derived neural activity on the scalp, is promising because it can be used safely and easily [10]. Therefore, this is expected to present sufficient opportunities for cognitive training.

Recently, we extended existing systems, such as the P300 speller [11], with our analytical methods in neurophysiology [12, 13] to develop the "Neurocommunicator®" (NC), which was designed as a practical BMI system (press release: 2010/03/29) [14, 15] (Fig. 16.2). The user focuses on one of eight pictograms (simplified drawings of medical care, nursing care, etc.) in a 3 × 3 matrix (excluding the center) on a monitor. After the system begins, each pictogram is pseudo-randomly intensified (flashed) at 8 Hz for several blocks, displaying four Japanese characters (e.g., "Ko-Re-Ka-Na"), which means "This one?" in English. An event-related potential (ERP) response is observed when the user intentionally reacts to a flashed target. Although ERP is a small potential change that reflects a temporal change in attention [16, 17], the real-time detection of the ERP response enables us to use it as a virtual switch (known as an "EEG switch").

Three core technologies (a-c) are highlighted by the NC. (a) The "Neurorecorder®" (fundamental hardware) consists of custom-made plastic headgear worn around the top of the head with a small wireless EEG amplifier attached to it, the five or eight electrodes connected to the amplifier via lead cables, and the shield cover to remove electrical noise. (b) The "Virtual Decision Function" is a basic analytical NC method for high-speed and high-accuracy decoding [12–14] that can shorten the stimulus presentation time (number of flashes). (c) The "Hierarchical Message Generation System" enables users to convey various messages quickly. Although the number of pictograms for a single choice was restricted to eight, the user could choose as many as 64/512 pictograms in 2/3 consecutive selections (8^2 or 8^3). The selected message was expressed by an artificial voice lip-synced by the CG avatar.

Although the main index of NC performance (the decoding accuracy of the target) is generally high in healthy volunteers as well as in many patients, such as patients with amyotrophic lateral sclerosis (ALS), it tends to be lower in the older and/or long-term bedridden patients. These observations imply that the ERP response is correlated with cognitive ability, and that ERR-based BMI might be an effective tool for monitoring and training cognitive abilities. For this reason, we focused on the feasibility of an ERP-based cognitive training system to prevent dementia in older individuals and began developing the "Neurotrainer®," which utilizes the EEG switch to conduct cognitive tasks (games) [18, 19].

Cognitive training using EEG-based BMI is effective as a social activity where the players are motivated by a competitive spirit to participate together with multiple participants, similar to various formes of sports. Therefore, as a candidate system/service that contributes to the prevention of dementia in older individuals who are at risk of deterioration in both cognitive and motor functions, we investigated the brain training competition called "bSports" (b standing for "brain") using an EEG switch. Here, we present our efforts to achieve this goal.

16.2 Experiment (1): Flashcard Games Based on an Oddball Task

16.2.1 Goal of Exp. 1

As the first step of developing the Neurotrainer, we focused on a visual discrimination task known as an "oddball task" [20], which has often been used in basic and clinical research on ERP. A classic oddball task consists of frequent stimuli ("standard") and less frequent stimuli ("target"). It has been observed that this deviant target reliably elicits a strong EPR across the skull, which usually occurs approximately 300 ms after stimulus presentation [16, 21]. The rules of this task are very simple and akin to "whack-a-mole." The odds of a correct answer are as high as 50%. Therefore, it was anticipated that many people would enjoy the oddball-type game, with the EEG switch offering at least 75% success rate (midway between a 50% and 100% chance of success).

16.2.2 Methods of Exp. 1

We recorded the ERP data from 11 healthy adult subjects ("players") during the oddball-type flashcard game, giving feedback to the players regarding success or failure [18] (Fig. 16.3). The procedure is as follows: The target was presented for 2 s

16 Maintenance and Improvement of Brain Health by the Brainwave-Based…

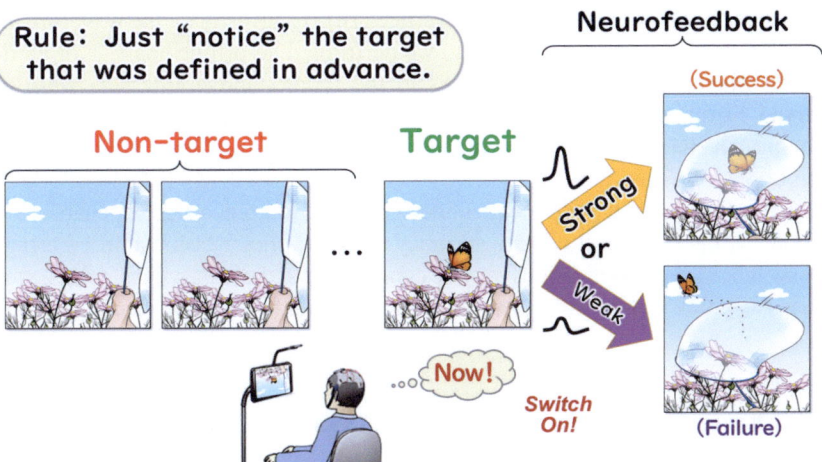

Fig. 16.3 This type of game was very simple even for older people

immediately before the start of each trial. Then, a specific non-target stimulus was presented several times, followed by the target being presented once. The rule of the game was that when the target appeared on the screen, the players only had to be conscious of whether "It appeared!" was in their mind. For example, in the case of a game that imitates the role of a batter playing baseball, after several presentations of an illustration showing "a pitch outside the strike zone" (non-target), an illustration showing "a pitch inside the strike zone" (target) appeared, and the batter was expected to swing only when the target was presented.

Success or failure was determined based on whether the EEG switch was activated for the target. Pattern-identification technology served as the basis for this determination, utilizing previously acquired calibration data. Specifically, we first performed linear discriminant analysis and created a discriminant model to calculate an average discriminant score of "+1" for the target and "-1" for the non-target. Discriminant scores were calculated immediately after presenting each stimulus. During the game session, we used the real-time decoding mode. The discriminant score for the presented target was compared with the average of the discriminant scores for non-targets presented multiple times before the target. If the discriminant score for the target was higher than the discriminant score for the non-target, it was assumed that "the EEG switch was activated" for the target and judged to be successful. Otherwise, it was judged as a "failure" as "the EEG switch was not activated." This process took only a second after the presentation of the target. Then, immediate feedback was given to the player; for the baseball game, it was labeled as a "big hit" for success or "a swing and a miss" for failure.

16.2.3 Results and Discussion of Exp. 1

All players succeeded in the oddball game with the EEG switch, with a high success probability of approximately 83% on average [18]. To investigate the waveform patterns contributing to such deciphering, we performed an offline analysis of all data for each player. Differential responses to the targets and non-targets were also observed. Stronger positive potentials were observed for the target than for the non-target in each channel, from approximately 250–400 ms on average after the presentation of the stimulus.

The results suggest the possibility of developing a Neurorecorder that enables high-speed decoding at higher than 75% accuracy for a two-choice task within 1 s using an EEG switch. These specifications may be comparable to those of typical games, where players can enjoy and improve their performance through repetitive training.

16.3 Experiment (2): Robot Sports Games Based on a Target Selection Task

16.3.1 Goal of Exp. 2

Next, we developed a game with more than just binar choices. In general, there is a trade-off between accuracy and speed in target selection for many objects [21]. Three tasks were performed to address this problem. First, we used a humanoid robot [22] to motivate players such that they could wait for several seconds (trials) for single choices until the decoding accuracy was sufficiently high. Second, based on the calibration data, we introduced a method that enabled us to decode the target with as few trials as possible when the response to the target was sufficiently strong. Third, we tested the Neurotrainer under competitive conditions between a pair of players, because this approach appeared effective in improving motivation through competitiveness and social interaction. In this experiment, we verified the effects of these attempts on player performance.

16.3.2 Methods of Exp. 2

The method of stimulus presentation in this experiment was identical to that used in the NC system. When the players saw "This one?" flash on the target picture card, they would think, "That's it!" This resulted in the pushing of the EEG switch to the target. However, instead of using an artificial voice synchronized with the lip movements of a CG avatar and speech balloon messages, the content of the selected picture card message was conveyed through the gestures of the robot avatar (Fig. 16.4). The robot was the commercial light-weight humanoid robot "KHR-3 HV"

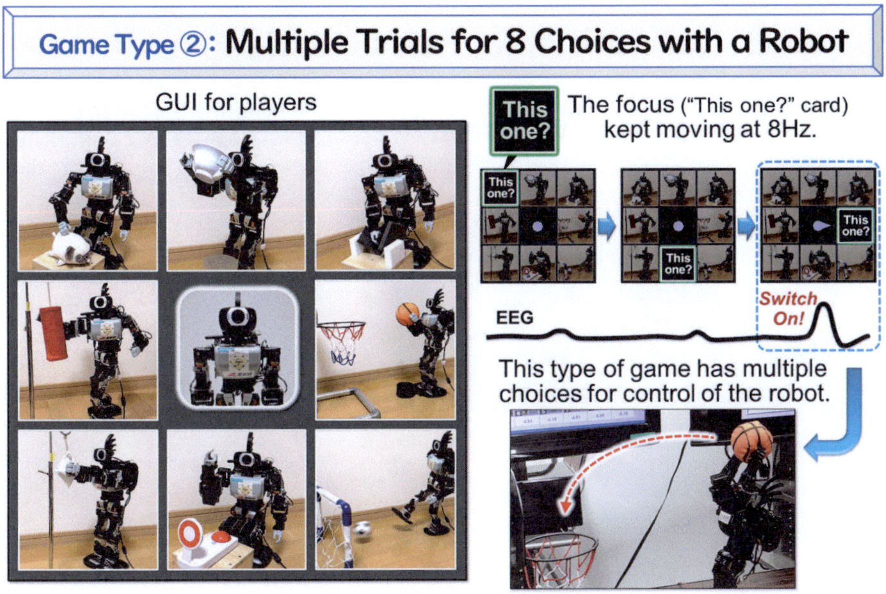

Fig. 16.4 This type of games can have multiple choices for robot control

(manufactured by Kondo Kagaku, inc.). This robot was used to express messages through NCs. The original reason for introducing the robot was that its gestures would enhance the sense of presence of unmovable patients. Specifically, KHR-3 HV exhibits a diverse range of anthropomorphic and naturalistic movements facilitated by its abundant degrees of freedom (22 axes) and robust servomotors [23]. However, we felt that the simple expression of gestures was not sufficient as a game for healthy players of bSports. Therefore, we selected and registered several sets of eight types of motions in each, mainly from popular sports, such as shooting a soccer ball, shooting a basketball, and breaking a tile in karate, which would be enjoyable to watch a robot perform [19].

An issue was that there were eight choices each time; therefore, the probability of success by chance was only 1/8 (12.5%). Unlike in the two-choice oddball game (chance level: 50%), it seemed impossible to achive more than 50% success in a single trial. However, decoding the target from multiple trials each time would impair the sense of speed. Therefore, we applied the virtual decision function (VDF) to the ERP data in this experiment. The VDF is a high-speed and high-accuracy decoding method [12, 13]. It is similar to how an election shows a surefire winner for one of the candidates; we can statistically guarantee which candidate will be elected when counting votes in an election. The VDF also quantitatively evaluates the response strength of the EEG switch based on the calibration data. We could then terminate the flash as soon as the target was accurately predicted.

Fifteen pairs of 30 players participated in the experiment (Fig. 16.5). Of these, 17 were aged over 60 years. Each pair played four different robot manipulation games

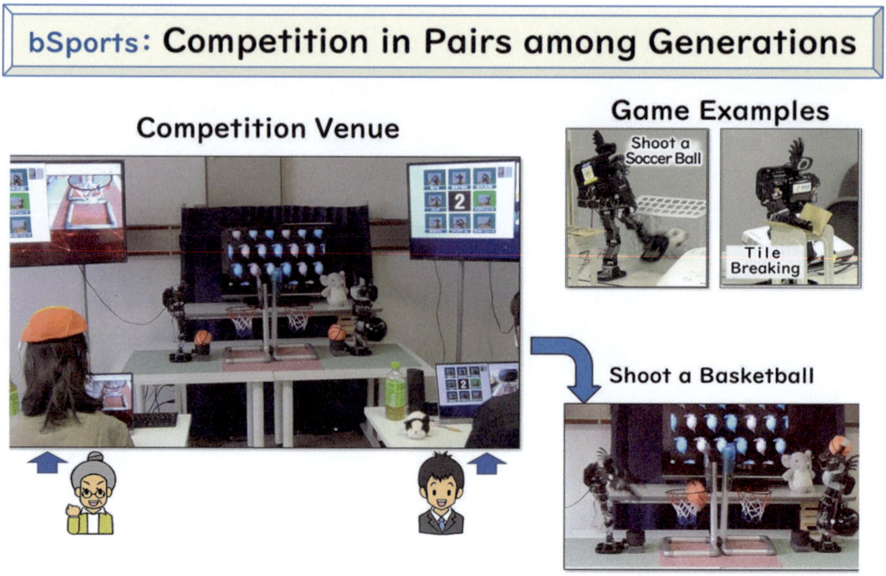

Fig. 16.5 In bSports, older people can compete on equal footing with young people

alternately, and the player with the most success was the winner. If both players succeeded in the same game, the winner was the player who finished the game in fewer trials. We focused on the success rate and number of trials for each player as the main experimental results.

16.3.3 Results and Discussion of Exp. 2

Fifteen pairs of healthy adults, including older people with a tendency for motor function degradation, were involved in a one-hour session of the four games [19]. Each player engaged the EEG switch to select one of the eight pictograms, thereby controlling their robot avatar to perform the desired action using the decoding method proposed by the VDF. Irrespective of age, all participants demonstrated proficient performance, with an average success rate of approximately 85% in 4.5 trials, corresponding to 4.5 s. The success rate provided by the VDF exceeded that of the traditional method, achieving a higher rate (85%) than the estimated 75% accuracy rate of traditional methods over an average of 4.5 trials, as indicated by the results obtained through cross-validation of the calibration data [19]. Alternatively, viewed from another perspective, the number of trials (4.5) required was fewer than the 7 trials estimated using traditional methods with an 85% accuracy level [19].

There were no significant differences in the success rate or number of trials related to age or sex. In addition, the players reported that they had fun playing bSports robot games in a competitive setting.

In summary, in Experiment 2, there was an improvement on the following three points compared to Experiment 1. (1) When increasing the number of choices from two to eight, one choice was made in approximately 5 s. (2) As a result of selection by EEGs, the humanoid robot's actions with a sense of realism increased its attractiveness. (3) It was confirmed that the older and young people could play against each other on equal footing. These results suggest that bSports is an effective cognitive training method and social activity in which people of all ages can participate.

16.4 Summary and Conclusions

In this chapter, we introduce the history of the bSports project, demonstrating that hands-free cognitive training using the EEG switch is possible even for older people [18, 19]. The Trail Making Test (TMT), which is famous as one of the tests that contribute to the early detection of MCI, aims to quantitatively evaluate cognitive function by connecting numbers and letters in order with a pencil and measuring the time taken [24]. Conventional cognitive tasks, such as the TMT, are limited because older people take longer to complete them [25–27]. However, it is difficult to evaluate pure cognitive function because of the possible influence of decreased motor function. In addition, if there is a change in performance owing to training, it is difficult to judge whether it is due to improvements in motor or cognitive function. The greatest advantage of bSports is that the brain-training games can be conducted using an EEG switch that does not depend on motor functions.

The first step is demonstrating that an EEG switch can work for various games. The original purpose was to demonstrate the feasibility of preventing or improving cognitive decline through repeated bSports experience. Therefore, through joint research with the University of Tsukuba Hospital, we are currently verifying the effects of repeated bSports experience on individuals who have undergone existing cognitive tests and have not yet shown signs of cognitive decline. Although this experiment is still in progress, we have already confirmed an improvement in game performance for five pairs of ten people, including older people.

We will continue to conduct such experiments in the future and will start to work on patients with a slight decline in cognitive function, specifically patients with MCI as well as long-term bedridden patients due to motor dysfunction. It would be better for them to use Neurotrainer continuously at home or in a daycare facility to help maintain and improve brain health. In addition, we would like to increase the opportunities for social interaction that transcend generations and disabilities by holding regular bSports tournaments, which may also lead to "town revitalization." If local companies sponsor prizes, this would motivate and excite citizens to participate. Furthermore, if more citizens participate, medical and nursing care costs for local governments will be reduced. Therefore, providing tax incentives to companies participating in the sports industry will also lead to returns in the long run, with corporate growth and increased tax revenue. Considering this, we believe that local governments can create a "virtuous cycle" by providing incentives to bSports participants through health points and other benefits.

Even those who are not previously familiar with bSports should know it is a core gear that will become a driving force to revitalize citizens, local governments, and companies, aiming to create cities where people and industries shine. The trend toward the practical use of bSports will contribute to achieving some of the Sustainable Development Goals (SDGs) adopted at the United Nations Summit in September 2015.

References

1. Fried LP, et al. Frailty in older adults: evidence for a phenotype. J Gerontol A Biol Sci Med Sci. 2001;56(3):M146–56.
2. Gauthier S, et al. World Alzheimer report 2021: journey through the diagnosis of dementia. London: Alzheimer's Disease International; 2021.
3. Petersen RC, Morris JC. Mild cognitive impairment as a clinical entity and treatment target. Arch Neurol. 2005;62(7):1160–3; discussion 1167.
4. Manly JJ, et al. Frequency and course of mild cognitive impairment in a multiethnic community. Ann Neurol. 2008;63(4):494–506.
5. Plassman BL, et al. Systematic review: factors associated with risk for and possible prevention of cognitive decline in later life. Ann Intern Med. 2010;153(3):182–93.
6. Ngandu T, et al. A 2 year multidomain intervention of diet, exercise, cognitive training, and vascular risk monitoring versus control to prevent cognitive decline in at-risk elderly people (FINGER): a randomised controlled trial. Lancet. 2015;385(9984):2255–63.
7. World Health Organization. Risk reduction of cognitive decline and dementia. Geneva: WHO Guidelines; 2019.
8. Donoghue JP. Connecting cortex to machines: recent advances in brain interfaces. Nat Neurosci. 2002;5(Suppl):1085–8.
9. Nicolelis MA. Brain-machine interfaces to restore motor function and probe neural circuits. Nat Rev Neurosci. 2003;4(5):417–22.
10. Wolpaw JR, et al. Brain-computer interfaces for communication and control. Clin Neurophysiol. 2002;113(6):767–91.
11. Farwell LA, Donchin E. Talking off the top of your head—toward a mental prosthesis utilizing event-related brain potentials. Electroencephalogr Clin Neurophysiol. 1988;70(6):510–23.
12. Hasegawa RP, Hasegawa YT, Segraves MA. Single trial-based prediction of a go/no-go decision in monkey superior colliculus. Neural Netw. 2006;19(8):1223–32.
13. Hasegawa RP, Hasegawa YT, Segraves MA. Neural mind reading of multi-dimensional decisions by monkey mid-brain activity. Neural Netw. 2009;22(9):1247–56.
14. Hasegawa RP, Nakamura Y. An attempt of speed-up of neurocommunicator, an EEG-based communication aid. In: Lecture Notes in Computer Science (LNCS), vol. 9947. Cham: Springer; 2016. p. 256–63.
15. Takada H, Yokoyama K. Bio-information for hygiene. Singapore: Springer; 2021.
16. Polich J. Updating P300: an integrative theory of P3a and P3b. Clin Neurophysiol. 2007;118(10):2128–48.
17. Sutton S, et al. Evoked-potential correlates of stimulus uncertainty. Science. 1965;150(3700):1187–8.
18. Hasegawa RP, et al. Development of the cognitive training system by the event-related potential. Trans Jpn Soc Kansei Eng. 2021;20(1):49–57.

19. Hasegawa RP, Takehara MS, Yamamoto HS. Feasibility of "bSports" as an EEG-based cognitive training -a report on competitive fighting using the robot avatar. Trans Jpn Soc Kansei Eng. 2021;20(3):221–31.
20. Squires NK, Squires KC, Hillyard SA. Two varieties of long-latency positive waves evoked by unpredictable auditory stimuli in man. Electroencephalogr Clin Neurophysiol. 1975;38(4):387–401.
21. Hasegawa RP, Nakamura Y. Properties of the event-related potentials during the target selection task—toward the development of the cognitive assessment system. Trans Jpn Soc Kansei Eng. 2020;19(1):89–96.
22. Hasegawa RP, Nakamura Y. An EEG-based communication aid that uses the Robot Avatar. In: Proceedings of the 8th International Workshop on Biosignal Interpretation (BSI2016), vol. P1–6; 2016. p. 76–7.
23. KHR-3—ROBOTS: Your Guide to the World of Robotics; https://robotsguide.com/robots/kondo. Accessed 1 Jan 2020.
24. Partington JE, Leiter RG. Partington's pathway tes. Psychol Ser Center J. 1949;1:9–20.
25. Davies AD. The influence of age on trail making test performance. J Clin Psychol. 1968;24(1):96–8.
26. Kennedy KJ. Age effects on trail making test performance. Percept Mot Skills. 1981;52(2):671–5.
27. Tombaugh TN. Trail making test a and B: normative data stratified by age and education. Arch Clin Neuropsychol. 2004;19(2):203–14.

Chapter 17
Wearable Systems Supporting Healthy Daily Life

Guillaume Lopez

Abstract The usage of wearable devices such as smartwatches and wireless earphones has spread increasingly in recent years. Their functionalities are to keep track of body movements, location, and biometric signals and to inform and entertain the wearer mainly by visual and auditory means. These functionalities are generally put together and provided to the user in the form of application software (appli) that often support fitness and sleep monitoring. This chapter presents two case studies that go beyond fitness and sleep analysis and show how Wearables will enable the creation of systems that support healthy daily life.

Keywords Wearables · Eating habits · Thermal comfort · Heatstroke · Behavior transformation

17.1 Introduction

Wearable devices (Wearables) are any kind of electronic device designed to be worn on the user's body, implying processing and communication capabilities. Their functionalities are mainly to keep track of movements, location, and biometric signals and to inform the wearer through displays, sound, vibration, etc. The variety of wearable devices, such as smartwatches and wireless earphones, has increased rapidly, and their usage spread in recent years. This chapter presents two case studies that go beyond fitness and sleep analysis and show how Wearables will enable the creation of systems that support healthy daily life.

G. Lopez (✉)
Department of Integrated Information Technology, College of Science and Engineering, Aoyama Gakuin University, Sagamihara, Kanagawa, Japan
e-mail: guillaume@it.aoyama.ac.jp

© The Author(s), under exclusive license to Springer Nature Singapore Pte Ltd. 2024, corrected publication 2024
T. Shiozawa et al. (eds.), *Gerontology as an Interdisciplinary Science*, Current Topics in Environmental Health and Preventive Medicine,
https://doi.org/10.1007/978-981-97-2712-4_17

17.2 Supporting Healthy Eating Habits by Wearable Sensors and IoT Devices

17.2.1 Relationship Between Obesity and Dietary Habits

Measures against obesity include moderate exercise and improvement of dietary habits. Regarding dietary habits, most of the measures focus on the content of meals, such as nutritional balance. However, as it has been shown that people who eat fast tend to be obese, dietary habits such as the number of times they chew are also strongly related to obesity. From the viewpoint of medicine and health as a countermeasure against obesity, it is known that the analysis of dietary habits is important. Still, most of the countermeasures focus mainly on the contents of meals, such as nutritional balance and intake/consumption of calories.

In recent years, there has been an increasing number of studies focusing on the monitoring and feedback of the number of chewing cycles and the appropriate amount of food for individuals. Mastication is the process of chewing food taken into the mouth [1]. Zhu et al. suggested that eating slowly may prevent the development of metabolic syndrome [2]. Furthermore, it has been clarified that there is a strong relationship between fast eating and obesity, as BMI (Body-Mass Index) tends to increase with faster eating speed [3]. Besides, it has been clarified that less frequent chewing leads to obesity and that frequent chewing before swallowing reduces subjective appetite [4]. In addition to eating speed, conversation during meals has also been shown to be related to health [5].

Commercially available wearable devices are capable of monitoring human activity levels in relation to calorie output. However, there is still no device that automatically detects eating behavior in a free environment.

17.2.2 Forefront of Technology to Quantify Eating Habits

Some eating behavior analysis and recognition methods use special devices. For example, there is a multi-sensor necklace, "NeckSense," to enable automatic monitoring of eating activity [6]. NeckSense consists of a proximity sensor that captures chewing behavior from a distance between the device and the jaw, an ambient light sensor that captures the eating behavior of carrying food to the mouth, and an inertial measurement unit (IMU) that captures the movement of the forward-leaning posture when eating. Mastication and eating episodes are detected from the features. There is also a research team that developed a glasses-type device for estimating the number of mastication [7]. A piezoelectric strain sensor is placed on the temple of glasses to capture the movement of the temporal muscle during mastication. The signal is continuously divided every few seconds. After estimating whether it is a mastication segment using machine learning, the number of mastication is calculated with 96% accuracy using a linear regression model for the mastication segment. However, we can point out the following three problems as the main problems present in many studies that have high performance, as described above.

- It is difficult to introduce because it is not a commercially available device that anyone can obtain.
- Meals are detected, but a detailed analysis of eating behavior cannot be extracted.
- The sensor signal segmentation method can extract continuous mastication, but it cannot segment each mastication, and the performance drops dramatically in a natural eating environment.

In the following sections, we discuss how and to what extent we can accurately quantify eating behavior in a natural eating environment and how we can provide feedback to help people change their eating behavior through research examples of our research group.

17.2.3 Estimation of Eating Behavior from Eating Sound Data in Natural Eating Environment

To quantify eating behavior from meal sounds collected in a natural daily life eating environment, the authors' research group focused on commercial wireless earphone microphones. They developed a robust mastication system to the natural eating environment and attempted to classify mastication, swallowing (food/drink), and speech with high accuracy.

Mastication data are naturally larger in eating behavior than speech and swallowing data. However, many data analysis methods using machine learning require not only a large amount of data but also a balanced data set. Oversampling methods can be used to balance evenly imbalanced datasets. The minority class oversampling technique called Synthetic Minority Oversampling TEchnique (SMOTE) is widely used to resolve the imbalanced amount of data per data label. Therefore, the authors adopt borderline-SMOTE, which is different from SMOTE and oversamples only minority class samples on the decision boundary.

In addition, it is effective to use a processing method called Features Engineering before performing machine learning so that the data can better represent each action. At that time, the authors devised the following two points, extracted a total of 958 feature values for each sample, and tried to improve the accuracy.

- Change the data distribution of some highly distorted features.
- Generation of new features: square root, square root, cube root, logarithm, reciprocal of each feature.

Using four machine learning models, we built models that learned chewing, food swallowing, drink swallowing, speaking, and other five eating behaviors. Table 17.1 summarizes the results of evaluating the generalization performance when constructing an RF model before and after oversampling using a test data set (25% saved before oversampling). Not only has the accuracy of all eating behaviors

Table 17.1 Comparison of generalization performance of machine learning models created with the original dataset, the SMOTE dataset, and the Borderline-SMOTE dataset (average)

Evaluation method	Average accuracy (F1 score)		
	W/o oversampling	SMOTE oversampling	Borderline SMOTE oversampling
	Basic features + RF model	Basic features + SVM model	Features engineering + RF model
Mastication	91%	77%	96%
Swallow (food)	6%	28%	81%
Swallow (drink)	9%	19%	76%
Speech	87%	91%	94%
Other	45%	51%	88%

improved, but the recognition accuracy of swallowing (food and beverages), which has been remarkably low so far, has also improved significantly and is approaching a practical level [8].

On the other hand, to quantify eating behavior in real time and with high accuracy in a natural eating environment, we are also investigating a method that can automatically extract (segmentation) sound segments representing eating behavior [9]. To evaluate the segmentation method and the generalization performance of the number of mastication, data labeling was done by multiple people, and only the labels that everyone agreed on were used to create a robust data set. We proposed a method for automatically segmenting bone-conducted sounds corresponding to detailed eating behaviors and evaluated the precision and recall of segments or detailed eating behaviors, and the results were 88.6% and 67.6%, respectively. In addition, we estimated the number of times of mastication for each user using the classification model proposed so far, and the accuracy of the number of times of mastication was estimated to be 84.5%. A related study showed that if the estimation accuracy of the number of chewing exceeds 80%, the user does not feel uncomfortable with the feedback of the number of chewing.

17.2.4 Feedback for Behavior Transformation

It is necessary to make use of the surrounding environment to encourage healthier behavior. Based on the accurate identification of mastication and utterance, by presenting in real-time during the meal the mastication count and conversation time, it is possible to increase the mastication number and improve speech awareness. In fact, it has been shown that real-time chewing frequency feedback on a smartphone screen increases chewing frequency [10].

Believing that it is necessary to use the surrounding environment to encourage self-restraint in humans, we proposed and prototyped a plate-driven system called "MealJammer" that inhibits eating without relying on self-restraint [11]. In

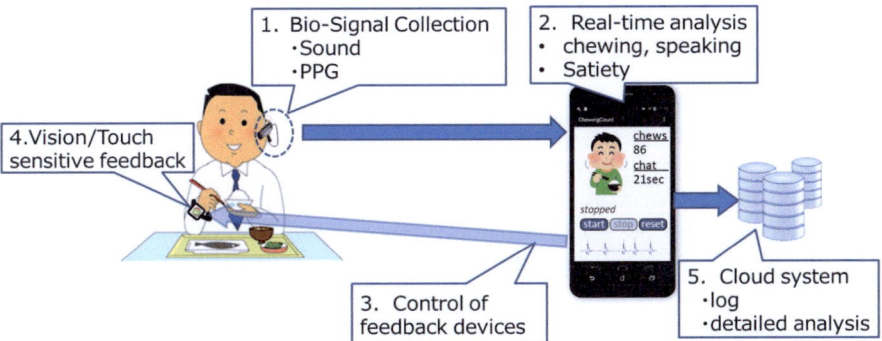

Fig. 17.1 Schematic of a system that supports healthy eating habits by wearable sensors and IoT devices

developing the proposed system, we investigated an "electromagnet" as an effective method for driving the plate, which enables stable feedback and physical information presentation. We assessed whether the system was capable of food inhibition. Ten healthy men and women in their twenties were asked to take meals on two different days, with and without MealJammer. We prepared two kinds of meals, chocolate and cucumber, and were asked to answer a questionnaire after the experiment. Besides, meal length was measured with a timer, the NASA-TLX questionnaire was used to evaluate the workload difference induced by MealJammer, and the SD method (semantic difference method) was used to evaluate the impression of the system. A Student's t test was performed to compare meal length with and without the proposed system. Though there was no significant difference for chocolate, a significant difference was found for cucumber.

We also proposed, developed, and evaluated a hydration promotion system called "HydReminder" to prevent water shortages (the name is a combination of "hydration" and "reminder") [12]. Ten men and women, from teens to grown-ups in their 50 s, participated in the experiment to identify the requirements of HydReminder. To measure the amount of water consumed in one rehydration and the time required for rehydration, they were asked to drink water five times on their own time. We measured the amount of water ingested using a digital scale each time rehydration was performed and compared the amount of water ingested and the frequency of water intake with and without HydReminder. Both the amount of water intake and frequency increased for eight out of ten persons. Regarding the frequency of water intake, it increased for eight out of ten persons but did not decrease for the other two persons.

Based on the above results and issues, if we build a system that can quantify eating habits and provide feedback that transforms into healthy eating behavior, as shown in Fig. 17.1, it will be possible to achieve healthy eating as a new countermeasure against obesity.

17.3 Wearable Sensor-Based Affect-Aware Intelligent Thermal Comfort Environments

17.3.1 Thermal Comfort and Health Risks

Thermal comfort provision is energy intensive. Building consumes 40% of the global energy, a large portion of which is solely spent on thermal comfort provision [13]. However, despite the dedicated energy resources, thermal comfort provision mechanisms (e.g., air conditioning units) are ineffective, and most building occupants complain about the lack of adequate thermal comfort [14, 15]. Thermal comfort is "the condition of mind that expresses satisfaction with the thermal environment and is assessed by subjective evaluation" [16]. Therefore, thermal comfort is a psychological sensation that varies from one person to another. Paradoxically, however, most thermal comfort provision technologies (e.g., air conditioning units) provide neutral thermal conditions to all occupants of the buildings. Unfortunately, this strategy is inefficient and has many well-known flaws highlighted in, e.g., [17]: First, a one-size-fits-all strategy cannot work well because of individual differences (e.g., age, gender, and physiological makeup) that influence how each person perceives thermal comfort [18]. Second, in reality, people prefer non-neutral conditions [19]. Third, achieving thermal neutrality is costly and necessitates immoderate energy consumption [13]. Fourth, only a few body parts (e.g., head, wrists, and feet) are mostly responsible for thermal comfort. For example, in uniform environmental conditions, when it is cold, a person's feet and hands feel colder than other body parts.

On the contrary, the cold environment does not affect the thermal sensation on the head, which usually feels warmer than the rest of the body and requires a relatively lower temperature to achieve satisfactory thermal comfort [20]. However, air conditioning units do not exclusively direct the heat to these crucial parts of the body. Instead, they inefficiently cool or warm an entire room regardless of the number of people available in the room. Lastly, current international thermal comfort standards are unambitious. They expect a mere 80% satisfaction rate [16], a scant performance rarely met in practice [15]. Recent awareness campaigns for a sustainable economy have led policymakers to enact energy reduction regulations. However, technological limitations seem to thwart this effort. For example, a mandatory energy-saving policy, which was introduced by the Japanese government, resulted in increased thermal dissatisfaction and a reduction in productivity [21]. Thus, delivering higher-quality thermal comfort at lower energy consumption is a conundrum that requires a paradigm shift in providing thermal comfort [22].

17.3.2 Human Factors-Based Thermal Comfort Provision

Current thermal comfort mechanisms consume a lot of energy, yet they provide inadequate thermal comfort. The author's research team is developing wearable devices-based technology to resolve this by mitigating the gap between the quality of the provided thermal comfort and the required energy consumption in an office environment. They propose to provide thermal comfort based on the person's physiological changes due to his surrounding environment. Indeed, humans maintain their body core temperature via a thermal regulation process that is mostly controlled by the brain's hypothalamus, which serves as the "thermostat" of the body. In a nutshell, the hypothalamus receives sensory inputs from thermo-receptors located in the skin, liver and skeletal muscles and initiates appropriate processes to keep the body's core temperature constant. For instance, when it is hot, the hypothalamus activates heat dissipating and body cooling mechanisms such as sweating and vasodilation. Conversely, when it is cold, the hypothalamus activates the thermogenesis mechanism and other mechanisms to reduce heat dissipation (e.g., cutaneous vasoconstriction and piloerection) [20].

Considering that thermal comfort is, by definition, a psychological sensation and depends on human thermoregulation, and given that thermoregulation activities induce detectable physiological changes, we made the hypothesize that thermal comfort state could be more accurately estimated based on the variation in the person's heart rate variability (HRV). The detected thermal comfort state could be used to fully automate indoor air conditioning based on people's thermal comfort level. Unlike existing thermal comfort estimation models, this approach would provide personalized thermal comfort to reflect each individual's thermal comfort expectations. Further, by this approach, it could be possible to significantly reduce the required thermal comfort provision energy by letting the indoor temperature drift away from thermal neutrality and adjust it only if people are about to feel thermally uncomfortable. Also, unlike existing systems that cool or warm an entire room, including its walls and furniture, regardless of the number of people present, the proposed approach could utilize a combination of centralized air conditioning and personalized thermal comfort provision systems in order to channel the thermal comfort to these parts of the body that are mostly responsible for the thermal discomfort. This may result in higher-quality thermal comfort and would require less energy.

We conducted experiments in thermal chambers and recorded human subjects' electrocardiogram (ECG). The ECG signal is used to compute HRV, which serves as an input to machine learning models that predict the comfort state (cold, neutral, and hot) of each subject. Thermal comfort is provided by creating a microclimate comfort zone around a person whereby the person's thermal comfort is estimated from the change in his/her heartbeat patterns. After that, appropriate utility functions could be used to select the most suitable thermal provision methods to meet everyone's thermal comfort needs at the lowest energy. These experiments conducted on human subjects lead to two important conclusions. First, the change in the thermal environment distinctively alters people's HRV, and heartbeats are more regular and less complex in comfortable environments and exhibit a more complex pattern in cold and hot environments. Second, in hot environments, neck cooling improves people's subjective perception of their thermal environments and leads to heartbeat patterns that are similar to what people would have had in a less warm environment [23, 24].

Then, we also investigated the possibility of predicting thermal comfort from people's HRV. The results strongly suggest that it is possible to design automated thermal controllers that predict people's comfort state based on their HRV. Two types of machine learning models were developed. The person-specific models were trained on the data of the same subject and evaluated using ten-fold cross-validation. It was found that person-specific models perform quite well (accuracy >95%) [25]. However, their usage is limited only to one specific person. On the contrary, the versatile generic models that could be used to predict the thermal comfort of any person do not work well (50% < accuracy<60%). Indeed, thermal comfort is expressed differently from one person to another [26]. Consequently, generic models cannot work well. Because thermal comfort varies from one person to another, a practical thermal comfort system based on human factors would only work if person-specific models are used. Unfortunately, this would be expensive and may not operate as well as expected because thermal comfort is dynamic and changes depending on unforeseeable factors. To tackle this issue, a practical and cost-effective calibration algorithm that derives an accurate and personalized affect (e.g., thermal comfort and stress) prediction machine learning model from physiological samples collected from a large population has proven efficient [26].

Finally, the influence of other types of stress on HRV cannot be omitted. Like thermal comfort, stress varies from one person to another. Consequently, although personal-specific models perform well in stress recognition, the generic models perform crudely. Studies have shown that although both thermal comfort and work stress affect HRV in an office environment, unless a person is both stressed and thermally dis-comfortable, most ephemeral changes in HRV are due to either work stress or thermal discomfort [27].

Furthermore, recently, Itao et al. proposed and developed the "Wearcon," a neck cooling and warming device [23]. In opposition to conventional air-conditioning systems, by cooling or warming the human body not through the air but directly, it improves the efficiency of the energy used, thus providing the possibility to reduce drastically the energy consumed at offices or home. For example, the above-presented algorithms for thermal comfort detection from HRV could be used in an intelligent environment that delivers efficient thermal comfort and improves the well-being of the occupants of the building. Simulations have shown that such architecture integrating individual thermal comfort provision devices (e.g., Wearcon) and conventional air-conditioning systems, the whole controlled by human factors, is efficient in reducing office energy consumption while ensuring a comfortable thermal environment for each person sharing the office space [28].

17.3.3 Application to Heatstroke Prevention

In addition to comfort issues, in recent years, the number of people suffering from heatstroke has been increasing. According to Japan's Ministry of Health, Labor, and Welfare statistics, the number of people dying from heatstroke in the 10 years to

2018 has almost doubled from the previous 10 years. According to the Fire and Disaster Management Agency, the number of people transported by ambulance has doubled from 5 years ago. Besides, according to the 2017 White Paper on Aging Society, more than seven million elderlies will live alone in 2025, but they are at risk of indoor heatstroke due to a shortage of helpers. Indeed, since 2010 about 80% of heatstroke deaths were elderly [29]. Avoiding this crisis has emerged as a social problem. It is necessary to take preventive measures not only for the elderly who are hard to feel the heat and infants who cannot cope with the heat by themselves but also for outdoor work. In fact, there are reports that heatstroke damage caused by continuous heat exposure and hypertension may be exacerbated unknowingly in a life exposed to an environment where temperature changes repeatedly. It is done based on the guidelines published by the International Labor Organization (ILO), Occupational Safety and Health Management (OSHMS); in Japan, the risk of the hot environment is estimated according to the risk assessment guidelines based on Article 28–2, Section 2 of the Industrial Safety and Health Act. To reduce the risk of heatstroke and prevent it from becoming more severe, it is necessary to measure body temperature, weight, and heart rate during work. Heart rate increases with dehydration. It is necessary to stop exposure to heat above 120 and to stop heat work when the temperature under the armpits exceeds 37.5 °C. Stop exposure to heat even if you have lost 1.5% of your weight before starting work.

Following the guidelines defined in the Risk Assessment Manual for Heat Stroke Prevention Measures, the author's team proposes to use wearable information and communication technology (ICT) to implement in society methods to protect the safety and security of the elderly living alone.

The challenge is to develop a heatstroke prevention system by utilizing the wearable devices that have already been developed to form a series of loops of human information sensing, processing, and action, to put it into practical use and to promote research in human informatics. With the proposed technology, in the thermoregulatory reaction during heat exposure and the mechanism of heatstroke, changes in the body such as autonomous nervous system state disorder or body temperature rise due to heat stress that the person does not notice can be detected at an early stage. By activating the cooling action, it supports regulatory functions such as vasodilation and sweating reaction, works directly to prevent body temperature rise, and can prevent aggravation. Such technology can provide a comprehensively appropriate thermal environment while realizing individual adaptation using human and environmental information. Besides, it can support both the mental and physical health of human beings. Therefore, we believe that it will significantly contribute to the "social implementation of safety and security technology based on human information and social information."

We propose an automatic integrated air-conditioning control system that reduces the biological load and discomfort of users in the space by quantitatively evaluating the thermal sensation of each individual in real-time based on energy-saving thermal human factors. Fig. 17.2 shows a schematic diagram of the proposed system, which aims to automatically control thermal comfort while minimizing power consumption. We aim for the social implementation of a wearable ICT system society

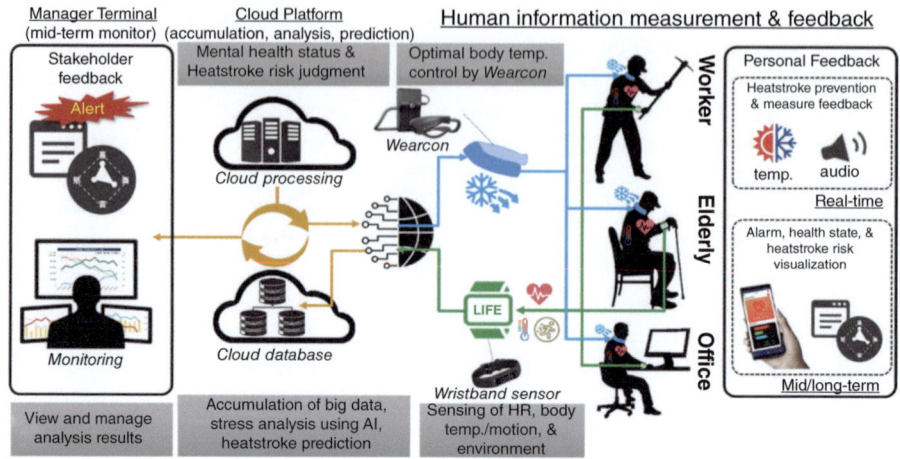

Fig. 17.2 Schematic Image of the Social Implementation of Wearable ICT System for Heatstroke Countermeasure Based on Human and Environment Information

that prevents heat stroke by clarifying the interrelationship between autonomic nerves and body temperature due to stress generated from the social environment and living environment, and based on human information such as heart rate, autonomic nerves, and body temperature, information processing by AI (general term for methods used in artificial intelligence technology) while utilizing processing and big data instructing physical condition management according to stress and discomfort, and performing sensor actions such as body temperature control according to each individual. Preliminary studies have shown that it is possible to automatically detect gradual changes in thermal discomfort level from vital signs measured continuously with a smartwatch, which is the key to predicting the risk of heatstroke [30] (Fig. 17.2).

References

1. Japan Society for Mastication: Message from the Japan Society for Mastication. http://sos-yaku.umin.jp/info/file/info01.pdf. Accessed: 29 Jan 2021.
2. Zhu B, et al. Association between eating speed and metabolic syndrome in a three-year population-based cohort study. J Epidemiol. 2015;25(4):332–6.
3. Ando Y, et al. ゆっくりとよく噛んで食べること」は肥満予防につながるか？ヘルスサイエンス ヘルスケア. 2008;8(2):51–63. (in Japanese).
4. Iwasaki M, et al. 成人期および高齢期における咀嚼回数と体格の関連. 口腔衛生学会雑誌. 2011;61(5):563–72. (in Japanese).
5. Mori H, et al. 女子学生の健康状況・生活習慣・食生活と小学生時の食事中の楽しい会話との関連. 日本家政学会誌. 2007;58(6):327–36. (in Japanese).

6. Chun KS, et al. Detecting eating episodes by tracking jaw-bone movements with a non-contact wearable sensor. ACM J. 2018;2(1):4.
7. Farooq M, et al. Segmentation and characterization of chewing bouts by monitoring temporalis muscle using smart glasses with piezoelectric sensor. IEEE J Biomed Health Inform. 2017;21(6):1495–503.
8. Nkurikiyeyezu K, et al. Classification of eating behaviors in unconstrained environments. In: Communications in computer and information science. Cham: Springer; 2021.
9. Kamachi H, Kondo T, Hossain T, Yokokubo A, Lopez G. Automatic segmentation method of bone conduction sound for eating activity detailed detection. In: The 9th International Workshop on Human Activity Sensing Corpus and its Application (HASCA 2021), in conjunction with UbiComp; 2021. Accessed 25 Sept 2021.
10. Lopez G, et al. Effect of feedback medium for real-time mastication awareness increase using wearable sensors. In: 12th International Joint Conference on Biomedical Engineering Systems and Technologies (HEALTHINF 2019); 2019.
11. Motokawa N, et al. MealJammer: 電磁石を用いた皿駆動型食事阻害システムの提案. IPSJ research report, entertainment computing (EC). 2020;56(14):1–2, 2188–8914. (In Japanese)
12. Motokawa N, et al. Hyd Reminder: ホルダー駆動型水分補給促進システム. In: The 193th meeting of human computer interaction special interest group (IPSJ SIG-HCI). Tokyo: IPSJ; 2021. (In Japanese).
13. Yang L, Yan H, Lam JC. Thermal comfort and building energy consumption implications—a review. Appl Ener. 2014;115(C):164–73. https://doi.org/10.1016/j.apenergy.2013.10.062.
14. International Facility Management Association. Temperature wars: savings vs. comfort. Houston: International Facilities Management Association; 2009. p. 1–7.
15. Karmann C, Schiavon S, Arens E. Percentage of commercial buildings showing at least 80% occupant satisfied with their thermal comfort. In: 10th Windsor Conference: rethinking comfort; 2018.
16. ASHRAE. Standard 55—2017 Thermal environmental conditions for human occupancy. Atlanta, GA: ASHRAE; 2017.
17. van Hoof J. Forty years of Fanger's model of thermal comfort: comfort for all? Indoor Air. 2008;18(3):182–201. https://doi.org/10.1111/j.1600-0668.2007.00516.x.
18. Wang Z, et al. Individual difference in thermal comfort: a literature review. Build Environ. 2018;138:181–93. https://doi.org/10.1016/j.buildenv.2018.04.040.
19. Brager GS, de Dear RJ. Thermal adaptation in the built environment: a literature review. Ener Build. 1998;27(1):83–96.
20. Arens E, Zhang H, Huizenga C. Partial—and whole-body thermal sensation and comfort—part I: uniform environmental conditions. J Therm Biol. 2006;31.1-2 SPEC. ISS:53–9. https://doi.org/10.1016/j.jtherbio.2005.11.027.
21. Tanabe S-i, et al. Thermal comfort and productivity in offices under mandatory electricity savings after the great East Japan earthquake. Archit Sci Rev. 2013;56(1):4–13. https://doi.org/10.1080/00038628.2012.744296.
22. Fergus Nicol J, Roaf S. Rethinking thermal comfort. Build Res Inform. 2017;45(7):711–6. https://doi.org/10.1080/09613218.2017.1301698.
23. Itao K, et al. Wearable equipment development for individually adaptive temperature-conditioning. J Jpn Soc Precis Eng. 2016;82(10):919–24. https://doi.org/10.2493/Jjspe.82.919.
24. Lopez G, et al. Effect of direct neck cooling on psychological and physiological state in summer heat environment. Mech Eng J. 2016;3(1):1–12. https://doi.org/10.1299/mej.15-00537.
25. Nkurikiyeyezu K, et al. Heart rate variability as a predictive biomarker for thermal comfort. J Ambient Intell Humaniz Comput. 2018;9(5):1465–77. https://doi.org/10.1007/s12652-017-0567-4.
26. Nkurikiyeyezu K, Yokokubo A, Lopez G. The influence of person-specific biometrics in improving generic stress predictive models. Sensors Materials. 2020;32(2(2)):703–22. https://doi.org/10.18494/SAM.2020.2650.

27. Nkurikiyeyezu K, Shoji K, Yokokubo A, Lopez G. Thermal comfort and stress recognition in office environment. In: The 12th International Joint Conference on Biomedical Engineering Systems and Technologies (HEALTHINF 2019), Prague, Czech Republic; 2019.
28. Lopez G, Aoki T, Nkurikiyeyezu K, Yokokubo A. Model for thermal comfort and energy saving based on individual sensation estimation. Sensors Materials. 2020;32(2(2)):693–702. https://doi.org/10.18494/SAM.2020.2635.
29. Kuzuya M. Heatstroke in older adults. Jpn Med Assoc J. 2013;56(3):193–8.
30. Hossain T, Kawasaki Y, Honda K, Nkurikiyeyezu K, Lopez G. The 4th international conference on activity and behavior computing. London: ABC; 2022.

Chapter 18
Experimental Understanding of the Flow Dynamics of Exhaled Air to Prevent Infection through Aerosol

Keiko Ishii, Yoshiko Ohno, Maiko Oikawa, and Noriko Onishi

Abstract Since several mass infections of COVID-19 have been reported in dense and enclosed environments, research efforts have been focused on aerosol transmission, which is a major mode of infection. Although the literature has traditionally focused on coughing and sneezing, it has become evident that avoiding exposure to the exhaled aerosols of asymptomatic infected persons is important for preventing infection. However, limited experimental cases have investigated ordinary exhaled air in real-world settings. Therefore, the authors of this study visualize the flow of exhaled aerosols with and without masks and face shields by simulating customer service in a beauty salon. Since the particle shape of the exhaled aerosol, which is problematic for infection, is similar to that of the particles released by e-cigarettes, light from a laser source scattered by e-cigarette smoke was captured by a camera. The results suggest that exhaled air flows according to certain rules under the influence of the temperature boundary layer on the human body surface. Therefore, it is evident that masks and face shields control the flow of exhaled air in such scenarios.

Keywords COVID-19 · Aerosol infection · Mucous membranes · Infectious disease

K. Ishii (✉)
Chuo University, Bunkyo, Japan
e-mail: k.ishii@mech.chuo-u.ac.jp

Y. Ohno · M. Oikawa · N. Onishi
Yamano College of Aesthetics, Tokyo, Japan

18.1 How Infections Spread Through the Air

18.1.1 Droplet Infection

People emit droplets of various sizes through their noses and mouths when coughing, sneezing, talking, breathing, and so on. The droplets exhaled by a person with an infectious disease contain the corresponding virus, and the infection is then transmitted through these droplets. These particles behave differently in the air depending on their size and the amount of liquid they contain, which in turn changes the characteristics of the mode of infection.

Large particles have a diameter greater than 5 μm and may be emitted by coughing, sneezing, or talking. Such particles fall within seconds of being emitted and do not pass through the gaps in the fabric of a mask. Therefore, wearing a mask when sneezing or coughing can eliminate the transmission of such droplets to others. However, patients who sneeze and cough frequently touch their eyes, nose, and mouth more often; thus, anything they touch with their virus-contaminated hands becomes contaminated. The risk of infection increases if a person in contact with such contaminated surfaces touches their mucous membranes or eats something. From this perspective, it is believed that the prevention of common colds and influenza could be achieved by frequent hand washing, disinfection of surfaces, and wearing a mask to avoid exposure to droplets released due to sneezing and coughing.

18.1.2 Airborne Infection

Airborne transmission refers to a route of infection that occurs when a person is infected with an airborne virus or bacteria. Diseases such as measles, chickenpox, and tuberculosis are airborne and are known to be highly contagious, even in dry conditions. For example, entering a space where an infected person has been could result in a high probability of infection.

In this type of transmission, the moisture of pathogens in the droplets released through sneezes, coughs, and breath spray dries out. The dried viral nuclei with a diameter of approximately 0.1 μm remain in the area for a long time. It is also spread by viruses falling to the floor, which are then blown about like dust by the wind or the movement of people. Because the virus remains highly contagious even when dry, children and others who have no history of the infection or have not been vaccinated are easily infected. It is difficult to prevent the spread of infection by isolation. Therefore, various vaccinations are available for children as young as 1 year of age.

18.1.3 Aerosol Infection

Until COVID-19 began to spread in 2020, it was believed that the main transmission route of such viruses was through droplets and airborne transmission. Further, it was believed that COVID-19 could not remain infectious in dry conditions, which is why it is not as infectious as measles and other viruses. Based on this assumption, wearing a mask to avoid exposure to droplets and washing and disinfecting hands frequently after touching something are sufficient to prevent the disease. However, despite this thoroughness, many infections were reported in enclosed, dense, and closely packed environments [1]. Therefore, aerosol infection has now become a growing concern.

Small particles with a diameter of less than 1 μm, which can be dispersed through conversation, coughing, or sneezing, can remain suspended in the air for several minutes to several hours [2]. Because aerosols contain water, their viruses are infectious when they have just been released. Due to their small size, they would normally dry out quickly, and the virus would not remain infectious in this state. However, the aerosols do not dry out quickly in a sealed, dense, and enclosed environment. This allows infectious particles to be inhaled by others or adhere to their mucous membranes.

Breathing also releases droplets; loud speaking or singing increases the number of particles released [3, 4]. The lesser the distance to an infected person (generally within 1–2 m), the higher the probability of infection, and the greater the distance, the lower the chance of infection [5, 6]. In particular, particles containing infectious viruses released from an infected person spend more time in the air in poorly ventilated environments or crowded rooms. Cases have been reported in many countries where infection occurred in such an environment despite maintaining a relatively great distance from the infected person [7, 8]. This has led to a common understanding of aerosol transmission.

One of the reasons for the explosive spread of COVID-19 is asymptomatic infection. In the past, infection could be prevented by staying away from symptomatic persons; however, it has been reported that asymptomatically infected persons with COVID-19 have a viral load that is not significantly different from that of those who are symptomatic [9]. Therefore, one must be wary of aerosol transmission from apparently healthy individuals, which may make preventing infection even more difficult.

According to Sayampanathan et al. [10], the infection rate of asymptomatic patients is said to be 3.85 times higher than that of symptomatic patients. This is thought to be due not only to an increase in the volume of bacteria carried by symptomatic patients but also to an increase in the number of droplets that are spread by symptomatic patients through sneezing and coughing, which increases the infection rate. Therefore, symptomatic patients tend to be more likely to transmit the virus than asymptomatic patients. Excessive fear of asymptomatic infection may also be a problem, which may cause social activities to stagnate.

In particular, because customer service, nursing care, and medical care are accompanied by speech, larger droplets are dispersed into the air than during normal breathing. After volatilization, the virus may become an aerosol and float in the air for a prolonged period of time. The risk is particularly high in the hairdressing, medical care, and nursing care industries, which are inseparable from human life because contact is a prerequisite, and it is difficult to refrain from speech. In the real world, it is necessary to reduce the risk of infection before carrying out activities. If we have an understanding of how human breath diffuses spatially, we can reduce the probability of aerosol exposure and reduce the risk of infection while providing customer service and medical services. Therefore, there has been extensive research on the flow dynamics of cough and sneeze droplets.

18.2 Literature Review of Experimental Methods for Flow by Exhalation or Coughing

Experimental visualization and discussion of the flow dynamics of the droplets released due to sneezing and coughing have been conducted since before the outbreak of COVID-19. Bourouiba et al. photographed such droplets with a high-speed camera and performed detailed physical analyses of their velocity and saliva deformation [11, 12]. Because sneezing and coughing are caused by the release of increased pressure in the oral cavity, the droplets are sprayed at high speeds over a wide area. It has been reported that these large droplets are sprayed at a speed of over 5 m per second, thus making it difficult to avoid them at a close range if no mask is used [13]. While the majority of such large droplets can be suppressed by masks, there are experimental reports of tiny droplets that can pass through the fabric and leak out of the mask surface due to coughing [14]. If the droplets are only suppressed with a handkerchief, the passage is significant, and the leaked droplets diffuse horizontally. In contrast, some passing microparticles are observed in nonwoven masks, but they do not spread as quickly and widely. Therefore, it is believed that droplet infection can be prevented by wearing masks. Thus far, although experiments have been conducted to visualize and measure airflow to accumulate knowledge on infection control, they have focused largely on sneezing and coughing. However, in order to prevent aerosol infection, it is necessary to determine the diffusion characteristics of the exhaled air not only during sneezing or coughing but also during normal breathing and conversation. The classical method of cough visualization is the Schlieren method, which is applied to observe thermal convection and shock waves based on the fact that the density of air affects the refractive index of light [15]. However, it is difficult to distinguish between thermal convection and droplets because thermal convection is also generated from the surface of the human body.

To verify the effectiveness of masks, Verma et al. used a smoke generator to perform smoke visualization and simulate exhalation; a mannequin wearing a mask was used to generate airflow by simulating a cough [14]. In conventional methods to visualize experimental exhalation, water vapor, smoke, and so on are used as tracer particles simulating exhaled aerosols, which are diffused through a mannequin. However, since exhaled air is considered to be affected by human body temperature, its movement may differ from actual exhaled aerosols.

Although several studies have been conducted to simulate the spread of exhaled air and coughing based on numerical calculations that also take into account the temperature and humidity of the human body, few measurements of real phenomena have been made based on experiments [16–18]. Although thermal mannequins have been used, they tend to be very costly and it is difficult to make them assume arbitrary postures or simulate conversation [19]. Yet, experiments using human subjects are rare due to the risks involved in relation to research ethics and the fact that reproducibility cannot be guaranteed. However, it is important to explore the flow dynamics of exhaled air emitted from an actual human body.

18.3 Methods of Experimental Flow Evaluation of Exhaled Aerosol

At present, limited research has focused on exhalation, and there have been no examples of visualization of situations in which face-to-face customer service is required. Therefore, the authors experimentally visualized the flow diffusion of the actual exhalation of a person speaking [20]. In particular, the effectiveness of wearing appropriate masks and face shields in preventing aerosol infection was verified by investigating how exhaled air diffuses when people are standing, sitting, looking up, and looking down to simulate scenes important in nursing care and hairdressing.

Droplets that contain viruses, such as saliva, are typically 1 μm ~ 1000 μm in diameter; viruses, such as novel coronaviruses, that are contained in these droplets are approximately 0.1 μm in diameter [21–23]. Droplets with a diameter of about 1000 μm are primarily generated by coughing and sneezing and common exhaled aerosols of about 5 μm or less. Therefore, e-cigarettes were used as tracer particles in this experiment. It has been reported that the liquid in e-cigarettes is mainly composed of glycerin and propylene glycol and becomes an aerosol with a particle size of 0.1–1 μm when heated [24, 25]. Therefore, using e-cigarettes as tracer particles can experimentally simulate the spread of viruses released from the human body through respiration while minimizing their impact on the human body. VAPORESSO Veco One Plus (Tradeworks) e-cigarettes were used. BI-SO liquid (BI-SO) consisting of 45% propylene glycol, 55% glycerin, and 5% flavorings was used for the tracer.

The experimental system consisted of a human, a laser (G3000, Kato Optical Labs) and a CMOS camera (FLIR GS3-U3-32S4). A laser sheet (wavelength: 532 nm) was injected into an arbitrary cross-section and the light scattered by the e-cigarette was captured by the camera. Assuming customer service with actual speech, a common Japanese greeting, "Onegai shimasu," was repeatedly uttered while images were taken. The camera exposure time was 20 ms and the shooting speed was 5 fps. The experiment was conducted in a beauty salon at Yamano College of Aesthetics in Hachioji City, Tokyo. The experiment was conducted from date, time, and outside air temperature and humidity were 17:00–19:30 h on September 29, 2020, with an external air temperature of 19.7–19.5 °C and humidity of 57–60% and from 16:00–18:00 h on September 30, 2020, with an external air temperature of 23.0–21.2 °C and humidity of 57–65%. After the room was ventilated, the windows and doors were closed and the air conditioning was not turned on. The angle of the face and posture of the subjects were varied and the diffusion of exhaled air both with and without a mask were examined. Masks made of non-woven fabric were used. Because the airflow was turbulent, at least three shots were taken in each case to ensure reproducibility. Postures were chosen to simulate customer service at a beauty salon, as well as similar face-to-face contact in nursing care, medical care, and so on.

18.4 Experimental Results

18.4.1 Exhalation in the Upright Position

First, the expiratory flow of an upright person was considered. Figure 18.1 shows the exhalation of a seated person without a mask. It can be observed that the exhaled aerosols diffuse downward in the form of a jet and then diffuse horizontally in a disordered manner.

Figure 18.2 shows the exhalation of an upright human wearing a mask. It can be observed that there is almost no flow of the exhaled aerosols towards the position of the laser sheet, i.e., forward. This is because the aerosols flow along the gap between the surface of the face and the mask in the opposite direction to that of the face. Because the exhaled air flowing in a downward direction from the front can be dramatically suppressed, wearing a mask reduces the risk of infection to the person in front. In such a scenario, exhaled air did not flow toward the front but rose along the surface of the human body and was released from the head. This is because the surface temperature of the human body is warmer than the air temperature, thus creating a "temperature boundary layer" on the surface of the human body. A

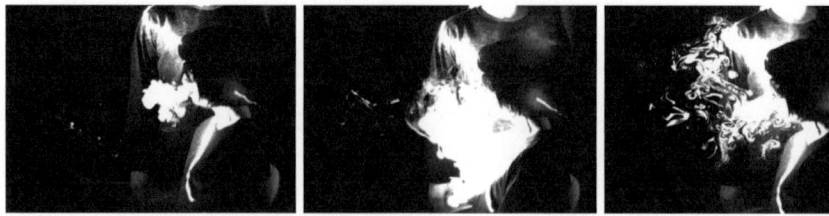

Fig. 18.1 Exhalation of an upright person

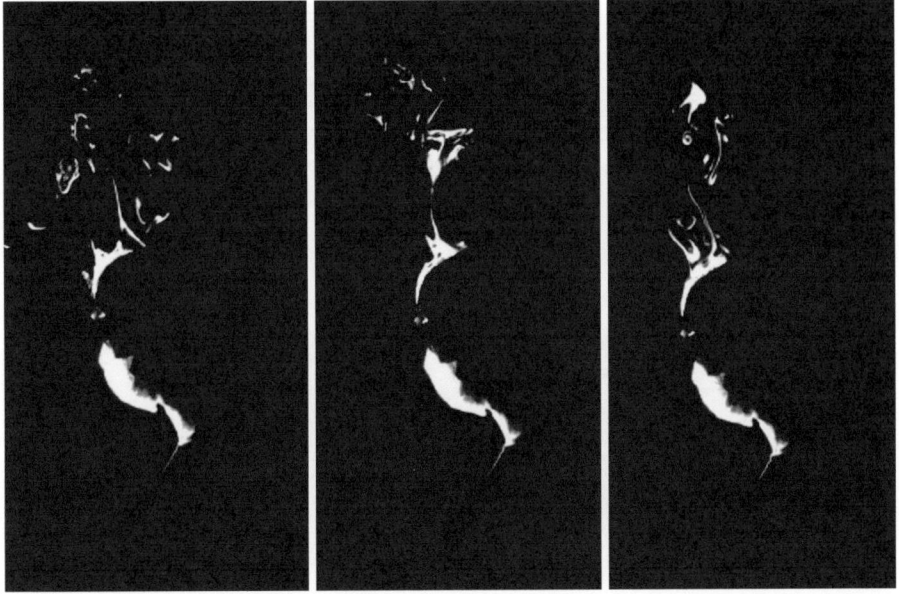

Fig. 18.2 Exhalation of an upright person wearing a mask

Fig. 18.3 Exhalation of a person looking diagonally down

temperature boundary layer is a region of rapid temperature change that is developed when another object at a different temperature is placed in the vicinity of an object. Here, natural convection occurs because the density of air changes with temperature. In other words, aerosols adhering to the surface of the human body due to the mask rise along the temperature boundary layer.

18.4.2 *Exhalation of a Person Looking Diagonally Down*

Figure 18.3 shows the exhalation of an unmasked person, assuming that they are a beautician performing a shampoo or other treatment and are positioned behind the customer. The client is seated, leaning their head on the shampoo stand. The

Fig. 18.4 Exhalation of a person wearing a mask and looking diagonally down

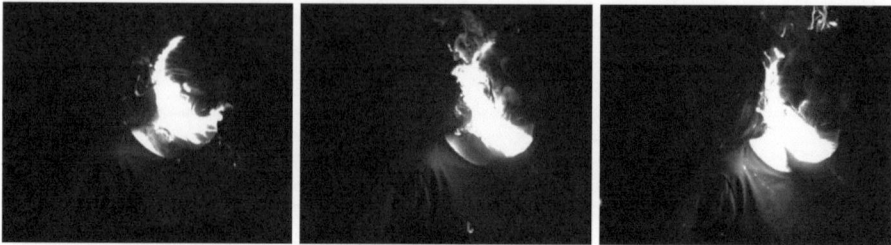

Fig. 18.5 Exhalation around the side of the head of a person wearing a mask and looking diagonally down

exhalation of the practitioner was directed downward and then stagnated in the space between the client and the shampoo table. Since the practitioner was not wearing a mask, there was a risk of prolonged exposure of the client's head and face to the aerosols from a low position.

Figure 18.4 shows the exhalation when the practitioner wears a mask. As expected, no exhalation jet was generated in front of them. It should be noted that although the permeation of exhaled air was also observed above the mask, there was nearly no air directed downward from the vicinity of the chin. This is because the temperature of the human body and exhaled air is higher than that of the surrounding atmosphere, which generates an updraft and suppresses the downward flow. In addition, exhaled air tends to flow upward more easily near the chin than near the nose because the mask tends to adhere more closely to the skin in this area.

Figure 18.5 shows the exhalation observed by the laser sheet incident from the temporal to the occipital area of the head. The exhaled air flowed along the mask and toward the occiput. It can be observed that the exhaled air flowed toward the occiput and slowly rose from the head if the mask was worn when the person was upright or reclining. This avoided exposing the person in front to the aerosols that were released directly from the mouth.

18.4.3 Exhalation of a Person Facing Diagonally Upward

Figure 18.6 shows the exhalation of a person facing obliquely upward without a mask. In this scenario, the exhalation flowed upward. It can be stated that the upward tilt of the face and the relative warmth of the exhaled air allowed it to rise easily. Figure 18.7 shows a person in the same position wearing a mask. In this scenario, the exhalation moved along the mask, rising closer to the forehead and tending to flow backward. Therefore, when placing a mask on a patient in a nursing or medical setting, it is important to recognize that the exhalation flows backward.

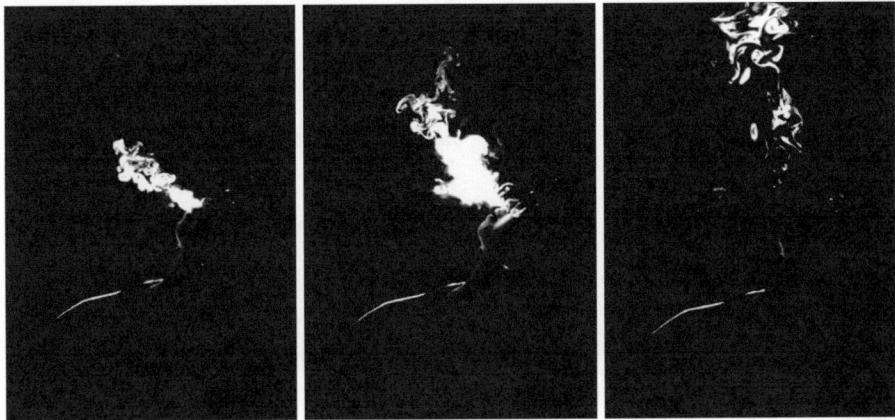

Fig. 18.6 Exhalation of a person looking obliquely upward

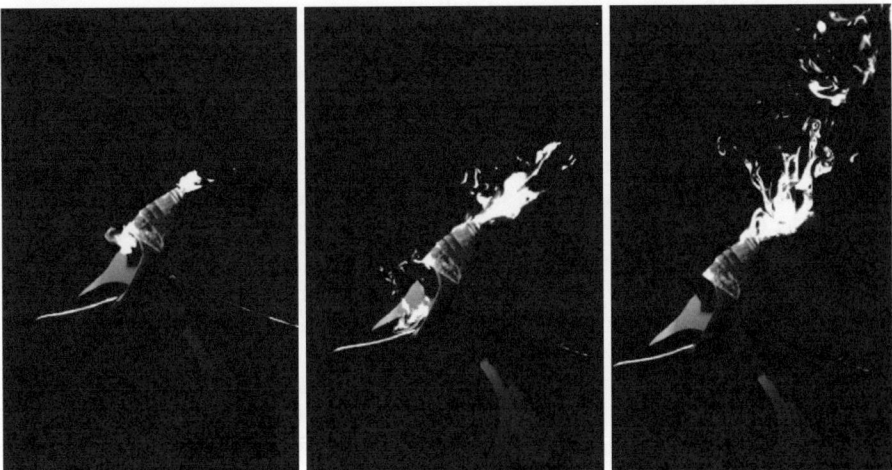

Fig. 18.7 Exhalation of a masked person looking obliquely upward

Fig. 18.8 Exhalation of a person looking upward

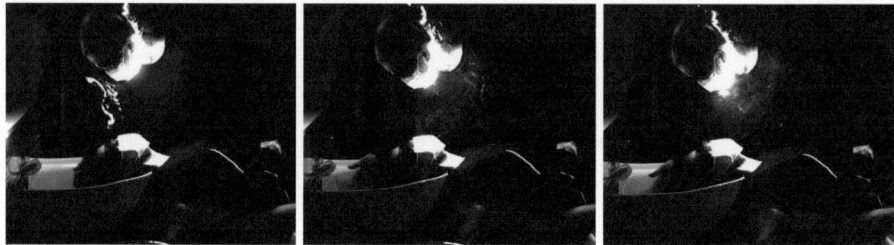

Fig. 18.9 Exhalation of a masked person looking upward

18.4.4 Exhalation of a Person Looking Upward

Figure 18.8 shows the exhalation of a person without a mask facing upward. This positional relationship occurs in various treatments performed on shampoo tables in both beauty salons and nursing homes. Here, the exhalation was directed upward, and the practitioner came into contact with it.

Figure 18.9 shows the exhalation when a mask was worn. The exhaled air leaking from the top and bottom edges of the mask was also directed upward and mingled with the breath emitted through the mask, resulting in a complex flow. Thus, it can be observed that there were no significant aerosol emissions from the e-cigarette from the surface of the mask, although they contributed to the generation of the airflow itself. The mask was able to suppress the forward flow of e-cigarette droplets as small as 0.1 μm, which suggests that particles as large as viral nuclei would not be able to pass through the mask.

Figure 18.10 shows the exhalation of the mask and face shield together. Exhaled air still flowed upward from the edges of the face shield.

18.4.5 Exhalation of a Person Looking Down

Figure 18.11 shows the expiratory flow of an unmasked person looking down. For the purpose of infection control, the client was not seated when this data was collected. As observed previously, the jet of exhaled air flowed downward in the

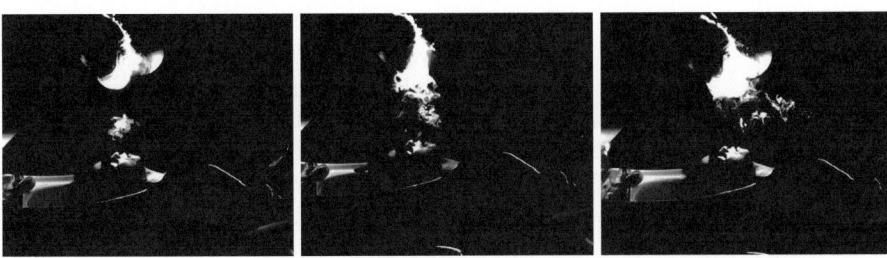

Fig. 18.10 Exhalation of a person wearing a mask and face shield looking upward

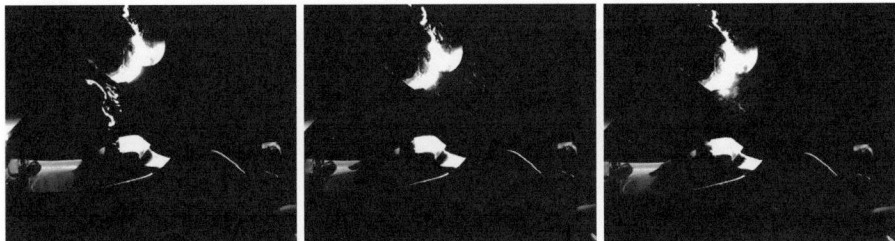

Fig. 18.11 Exhalation of a person wearing a mask facing up

Fig. 18.12 Exhalation of a person facing up

absence of the mask. It is evident that the exhaled air jet then remained in the vicinity of the shampoo table and was then diffused. The exhaled aerosols were still drifting in the air after several minutes.

Figure 18.12 shows the exhaled air when a mask is worn. Unlike the previous case, the exhaled air leaking from the mask fell downward. In such a scenario, the aerosol leaking from the mask leaves the thermal boundary layer of the human body and diffuses randomly. The temperature boundary layer generated in a material at a higher temperature than that of the ambient atmosphere is thinner at the bottom due to the effect of upward flow along the object's surface. Therefore, when the object faces down, the exhaled air leaking from the mask can easily leave the temperature boundary layer and diffuse randomly in the surrounding air.

Figure 18.13 shows the exhalation when a face shield was worn in addition to the mask. In this scenario, the exhaled air flowed upward along the head and did not flow downward. It is conceivable that the structure of the face shield alters the flow and that the exhaled air is reattached to the temperature boundary layer and warmed, thereby encouraging the exhaled air to rise. Consequently, when approaching a person at a lower level, it is effective to use a mask and face shield together. Although wearing a face shield was often reported as ineffective in preventing infectious diseases in Japan in 2020, the above experiment demonstrates its effectiveness in preventing exhaled air from being directed downward when looking down (Figs. 18.14 and 18.15).

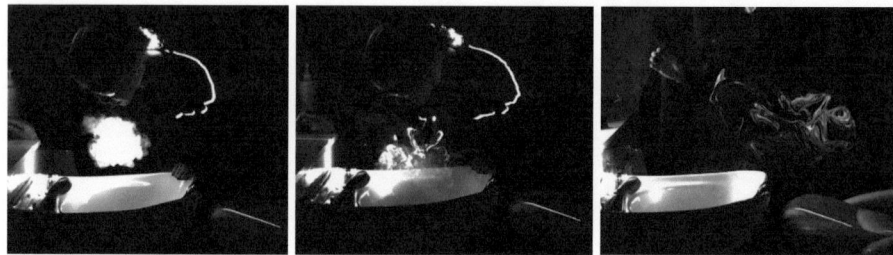

Fig. 18.13 Exhalation of a person looking down

Fig. 18.14 Exhalation of a person wearing a mask facing down

Fig. 18.15 Exhalation of a person facing downwards wearing a mask and face shield

18.5 Conclusion

In response to the outbreak of COVID-19, we visualized actual human exhalation to obtain guidelines for social activities, such as customer service, while minimizing the risk of infection. Based on our findings, several valuable conclusions can be drawn, which are as follows:

- If the exhaled air is relatively far from the body, the aerosols diffuse randomly.
- When wearing a mask, the exhaled air adheres to the human body, and the aerosols remain within the temperature boundary layer. When the person is in a slumped position, facing forward or looking upward, the aerosol flows along the body, rises, and is emitted from the head. This avoids direct exhalation towards a person positioned in front.
- When facing downward, aerosols are more likely to leave the human body due to the thin temperature boundary layer, and exhaled air is more likely to diffuse randomly. Therefore, exhalation is not directed downward when working with a person wearing a face shield and looking downward.
- When the mask is facing upward, the aerosols leaking from the upper portion of the mask tend to interfere with the exhaled air being emitted through the mask surface, which is then agitated and flows in a complex manner. In addition, exhaled air tended to flow from the edge of the mask to the rear. Therefore, care should be taken when establishing contact with a masked person from the rear. Even when a mask is used in combination with a face shield, there is a tendency for the aerosols to move upward around the edges of the face shield.
- Masks made of non-woven fabric can effectively suppress aerosols jetting to the front.
- Wearing a mask dramatically alters the movement of aerosols, which may be effective in preventing infection.
- If the aerosols are released from the head, the distance traveled will be longer and the aerosol will dry out, thereby reducing infectivity.

References

1. Jang S, Han SH, Rhee JY. Cluster of coronavirus disease associated with fitness dance classes, South Korea. Emerg Infect Dis. 2020;26:1917–20.
2. Wang CC, Prather KA, Sznitman J, Jimenez JL, Lakdawala SS, Tufekci Z, Marr LC. Airborne transmission of respiratory viruses. Science. 2021;373:eabd9149.
3. Stadnytskyi V, Bax CE, Bax A, Anfinrud P. The airborne lifetime of small speech droplets and their potential importance in SARS-COV-2 transmission. Proc Natl Acad Sci. 2020;117:11875–7.
4. Alsved M, Matamis A, Bohlin R, Richter M, Bengtsson P-E, Fraenkel CJ, Medstrand P, Löndahl J. Exhaled respiratory particles during singing and talking. Aerosol Sci Technol. 2020;54:1245–8.
5. Position paper on aerosols and SARS-COV-2. GAeF. https://www.info.gaef.de/position-paper. Accessed 19 Oct 2022.

6. Han ZY, Weng WG, Huang QY. Characterizations of particle size distribution of the droplets exhaled by sneeze. J R Soc Interface. 2013;10:20130560.
7. Toyokawa T, Shimada T, Hayamizu T, et al. Transmission of SARS-COV-2 during a 2-h domestic flight to Okinawa, Japan, march 2020. Influenza Other Respir Viruses. 2021;16:63–71.
8. Katelaris AL, Wells J, Clark P, Norton S, Rockett R, Arnott A, Sintchenko V, Corbett S, Bag SK. Epidemiologic evidence for airborne transmission of SARS-COV-2 during church singing, Australia, 2020. Emerg Infect Dis. 2021;27:1677–80.
9. Ra SH, Lim JS, G-un K, Kim MJ, Jung J, Kim S-H. Upper respiratory viral load in asymptomatic individuals and mildly symptomatic patients with SARS-COV-2 infection. Thorax. 2020;76:61–3.
10. Sayampanathan AA, Heng CS, Pin PH, Pang J, Leong TY, Lee VJ. Infectivity of asymptomatic versus symptomatic COVID-19. Lancet. 2021;397:93–4.
11. Scharfman BE, Techet AH, Bush JW, Bourouiba L. Visualization of sneeze ejecta: steps of fluid fragmentation leading to respiratory droplets. Exp Fluids. 2016;57:1–9.
12. Bourouiba L, Dehandschoewercker E, Bush JWM. Violent expiratory events: on coughing and sneezing. J Fluid Mech. 2014;745:537–63.
13. Wang H, Li Z, Zhang X, Zhu L, Liu Y, Wang S. The motion of respiratory droplets produced by coughing. Phys Fluids. 2020;32:125102.
14. Verma S, Dhanak M, Frankenfield J. Visualizing the effectiveness of face masks in obstructing respiratory jets. Phys Fluids. 2020;32:061708.
15. Tang JW, Liebner TJ, Craven BA, Settles GS. A Schlieren optical study of the human cough with and without wearing masks for aerosol infection control. J R Soc Interface. 2009;6:S727–36.
16. Dbouk T, Drikakis D. On coughing and airborne droplet transmission to humans. Phys Fluids. 2020;32:053310.
17. Chaudhuri S, Basu S, Kabi P, Unni VR, Saha A. Modeling the role of respiratory droplets in COVID-19 type pandemics. Phys Fluids. 2020;32:063309.
18. Dbouk T, Drikakis D. On respiratory droplets and face masks. Phys Fluids. 2020;32:063303.
19. Melikov A, Kaczmarczyk J. Measurement and prediction of indoor air quality using a breathing thermal manikin. Indoor Air. 2007;17:50–9.
20. Ishii K, Ohno Y, Oikawa M, Onishi N. Relationship between human exhalation diffusion and posture in face-to-face scenario with utterance. Phys Fluids. 2021;33:027101.
21. Yang S, Lee GWM, Chen CM, Wu CC, Yu KP. The size and concentration of droplets generated by coughing in human subjects. J Aerosol Med. 2007;20:484–94.
22. Chao CYH, Wan MP, Morawska L, et al. Characterization of expiration air jets and droplet size distributions immediately at the mouth opening. J Aerosol Sci. 2009;40:122–33.
23. Xie X, Li Y, Sun H, Liu L. Exhaled droplets due to talking and coughing. J R Soc Interface. 2009;9:S703–14.
24. Oldham MJ, Zhang J, Rusyniak MJ, Kane DB, Gardner WP. Particle size distribution of selected electronic nicotine delivery system products. Food Chem Toxicol. 2018;113:236–40.
25. Belka M, Lizal F, Jedelsky J, Jicha M, Pospisil J. Measurement of an electronic cigarette aerosol size distribution during a puff. EPJ Web Conferences. 2017;143:02006.

Correction to: Gerontology as an Interdisciplinary Science

Tomoki Shiozawa, Hiromi Hirata, Takashi Inoue, Dominika Kanikowska, and Hiroki Takada

Correction to:
T. Shiozawa et al. (eds.), *Gerontology as an Interdisciplinary Science*, Current Topics in Environmental Health and Preventive Medicine,
https://doi.org/10.1007/978-981-97-2712-4

This book was inadvertently published without chapter abstracts and keywords. It has now been updated.

In addition to this, in Chapter 9, an email address has been added to the author, Dr. Takeshita takeshita_kei@mac.co.

The updated version of this book can be found at https://doi.org/10.1007/978-981-97-2712-4

If you have any concerns about our products,
you can contact us on
ProductSafety@springernature.com

In case Publisher is established outside the EU,
the EU authorized representative is:
**Springer Nature Customer Service Center GmbH
Europaplatz 3, 69115 Heidelberg, Germany**

Printed by Libri Plureos GmbH
in Hamburg, Germany